ice
Institution of Civil Engineers

publishing

DESIGNERS' GUIDES TO THE EUROCODES

DESIGNERS' GUIDE TO EUROCODE 9: DESIGN OF ALUMINIUM STRUCTURES
EN 1999-1-1 AND -1-4

TORSTEN HÖGLUND
Royal Institute of Technology, KTH, Stockholm, Sweden

PHILIP TINDALL
Hyder Consulting, London, UK

Series editor
Haig Gulvanessian CBE

Published by ICE Publishing, 40 Marsh Wall, London E14 9TP

Full details of ICE Publishing sales representatives and distributors can be found at:
www.icevirtuallibrary.com/info/printbooksales

Eurocodes Expert

Structural Eurocodes offer the opportunity of harmonised design standards for the European construction market and the rest of the world. To achieve this, the construction industry needs to become acquainted with the Eurocodes so that the maximum advantage can be taken of these opportunities.

Eurocodes Expert is an ICE and Thomas Telford initiative set up to assist in creating a greater awareness of the impact and implementation of the Eurocodes within the UK construction industry.

Eurocodes Expert provides a range of products and services to aid and support the transition to Eurocodes. For comprehensive and useful information on the adoption of the Eurocodes and their implementation process please visit our website or email eurocodes@thomastelford.com

www.icevirtuallibrary.com

A catalogue record for this book is available from the British Library

ISBN 978-0-7277-5737-1

© Thomas Telford Limited 2012

ICE Publishing is a division of Thomas Telford Ltd, a wholly-owned subsidiary of the Institution of Civil Engineers (ICE).

Permission to reproduce extracts from EN 1990 and EN 1999 is granted by BSI. British Standards can be obtained in PDF or hard copy formats from the BSI online shop: www.bsigroup.com/Shop or by contacting BSI Customer Services for hardcopies only: Tel: +44 (0)20 8996 9001, Email: cservices@bsigroup.com.

Associate Commissioning Editor: Jennifer Barratt
Production Editor: Imran Mirza
Market Development Executive: Catherine de Gatacre

Typeset by Academic + Technical, Bristol
Printed and bound by CPI Group (UK) Ltd, Croydon, CR0 4YY

DESIGNERS' GUIDE TO EUROCODE 9: DESIGN OF ALUMINIUM STRUCTURES
EN 1999-1-1 AND -1-4

Eurocode Designers' Guide series

Designers' Guide to Eurocode: Basis of Structural Design. EN 1990. Second edition. H. Gulvanessian, J.-A. Calgaro and M. Holický. 978-0-7277-4171-4. Published 2012.

Designers' Guide to Eurocode 1: Actions on Buildings. EN 1991-1-1 and -1-3 to -1-7. H. Gulvanessian, P. Formichi and J.-A. Calgaro. 978-0-7277-3156-2. Published 2009.

Designers' Guide to Eurocode 1: Actions on Bridges. EN 1991-1-1, -1-3 to -1-7 and EN 1991-2. J.-A. Calgaro, M. Tschumi and H. Gulvanessian. 978-0-7277-3158-6. Published 2010.

Designers' Guide to EN 1991-1.4. Eurocode 1: Actions on Structures, General Actions. Part 1-4 Wind actions. N. Cook. 978-0-7277-3152-4. Published 2007.

Designers' Guide to Eurocode 2: Design of Concrete Structures. EN 1992-1-1 and EN 1992-1-2 General rules and rules for buildings and structural fire design. R.S. Narayanan and A.W. Beeby. 978-0-7277-3105-0. Published 2005.

Designers' Guide to EN 1992-2. Eurocode 2: Design of Concrete Structures. Part 2: Concrete bridges. C.R. Hendy and D.A. Smith. 978-0-7277-3159-3. Published 2007.

Designers' Guide to Eurocode 3: Design of Steel Buildings. EN 1993-1-1, -1-3 and -1-8. Second edition. L. Gardner and D. Nethercot. 978-0-7277-4172-1. Published 2011.

Designers' Guide to EN 1993-2. Eurocode 3: Design of Steel Structures. Part 2: Steel bridges. C.R. Hendy and C.J. Murphy. 978-0-7277-3160-9. Published 2007.

Designers' Guide to Eurocode 4: Design of Composite Steel and Concrete Structures. EN 1994-1-1. Second edition. R.P. Johnson. 978-0-7277-4173-8. Published 2012.

Designers' Guide to EN 1994-2. Eurocode 4: Design of Composite Steel and Concrete Structures. Part 2 General rules for bridges. C.R. Hendy and R.P. Johnson. 978-0-7277-3161-6. Published 2006.

Designers' Guide to Eurocode 5: Design of Timber Buildings. EN 1995-1-1. A.J. Porteous and P. Ross. 978-0-7277-3162-3. Forthcoming: 2012.

Designers' Guide to Eurocode 6: Design of Masonry Structures. EN 1996-1-1. J. Morton. 978-0-7277-3155-5. Published 2012.

Designers' Guide to Eurocode 7: Geotechnical Design. EN 1997-1 General rules. R. Frank, C. Bauduin, R. Driscoll, M. Kavvadas, N. Krebs Ovesen, T. Orr and B. Schuppener. 978-0-7277-3154-8. Published 2004.

Designers' Guide to Eurocode 8: Design of Structures for Earthquake Resistance. EN 1998-1 and EN 1998-5. General rules, seismic actions, design rules for buildings, foundations and retaining structures. M. Fardis, E. Carvalho, A. Elnashai, E. Faccioli, P. Pinto and A. Plumier. 978-0-7277-3348-1. Published 2005.

Designers' Guide to Eurocode 8: Design of Bridges for Earthquake Resistance. EN 1998-2 and -5. M.N. Fardis, B. Kolias and A. Pecker. Forthcoming: 2012.

Designers' Guide to Eurocode 9: Design of Aluminium Structures. EN 1999-1-1 and -1-4. T. Höglund and P. Tindall. Published 2012.

Designers' Guide to EN 1991-1-2, EN 1992-1-2, EN 1993-1-2 and EN 1994-1-2. T. Lennon, D.B. Moore, Y.C. Wang and C.G. Bailey. 978-0-7277-3157-9. Published 2007.

www.icevirtuallibrary.com
www.eurocodes.co.uk

Preface

General

EN 1999 applies to the design of buildings and civil engineering structures, or parts thereof, using 'aluminium'.

In the context of EN 1999, the term 'aluminium' refers to specific listed aluminium alloys.

This guide covers EN 1999-1-1 ('Eurocode 9: Design of aluminium structures – Part 1-1: General structural rules') and EN 1999-1-4 ('Eurocode 9, Design of aluminium structures – Part 1-4: Cold-formed structural sheeting').

It is noted that EN 1999-1-1 covers all structural applications, and, unlike EN 1993 (steel structures), there are not separate parts for bridges, towers and crane supports.

Material selection, all main structural elements and joints are covered within Part 1-1 of Eurocode 9.

Layout of this guide

The Introduction and Chapters 1–8 of this guide are numbered to reflect the corresponding section number of EN 1999-1-1. Chapter 9 of this guide covers the appendices of EN 1999-1-1, and Chapter 10 covers EN 1999-1-4 ('Cold-formed structural sheeting').

All cross-references in this guide to sections, clauses, subclauses, paragraphs, annexes, figures, tables and expressions of EN 1999-1-1 and EN 1999-1-4 are in italic type, which is also used where text from these two parts of Eurocode 9 has been directly reproduced. EN 1999-1-1 clauses cited in this guide are highlighted in the margin for ease of reference.

Commentary

EN 1999 has, along with all other Eurocodes, been produced over a number of years by experts from many countries. While EN 1999 has drawn material from previous national standards, including BS 8118, it is essentially a new document. Since publication in 2007, a number of errors have been identified and amendments and corrigenda issued to implement changes identified as necessary. This guide is based on EN 1999-1-1+A1+A2, and EN 1999-1-4+A1.

Wherever possible, the clauses and layout of Eurocode 9 have been written to mirror corresponding provisions in Eurocode 3. This has been done in an attempt to make it easier for designers switching from one material to another. However, it should always be remembered that aluminium is a very different material to steel. Aluminium has many benefits and much greater flexibility in product form, but additional specific design checks are needed that a steel designer might not anticipate.

Acknowledgements

The authors have benefited enormously from discussions within committee meetings and drafting panels for the production and maintenance of Eurocode 9. We are grateful to all of the experts who have participated in the production of the Eurocode.

H. Gulvanessian CBE
T. Höglund
P. Tindall

Contents

Designers' Guide to Eurocode 9: Design of Aluminium Structures
ISBN 978-0-7277-5737-1

ICE Publishing: All rights reserved
http://dx.doi.org/10.1680/das.57371.001

Introduction

The material in this introduction relates to the foreword to the European standard EN 1999-1-1, 'Eurocode 9: Design of aluminium structures – Part 1-1: General structural rules'.

The following aspects are covered

- the background of the Eurocode programme
- the status and field of application of Eurocodes
- the national standards implementing Eurocodes
- the links between Eurocodes and product-harmonised technical specifications (ENs and ETAs)
- additional information specific to EN 1999-1-1.

Background of the Eurocode programme

Work began on the set of structural Eurocodes in 1975, although work on Eurocode 9 did not start until 1990.

The first drafts of some of the Eurocodes (ENVs) started to appear in the mid-1980s. The fragmented nature and multiple parts of many Eurocodes, many of which were not published until much later, meant that the drafts were not readily usable for many applications.

ENV 1999-1-1 was published as a draft in 1998. Several countries carried out extensive calibration checks, and these checks gave rise to comments that were taken into account in the drafting of EN 1999-1-1, which was published in 2007.

The original, and unchanged, main grouping of the Eurocodes comprises ten standards each one generally comprising a number of parts. The ten standards are:

- EN 1990, 'Eurocode: Basis of structural design'
- EN 1991, 'Eurocode 1: Actions on structures'
- EN 1992, 'Eurocode 2: Design of concrete structures'
- EN 1993, 'Eurocode 3: Design of steel structures'
- EN 1994, 'Eurocode 4: Design of composite steel and concrete structures'
- EN 1995, 'Eurocode 5: Design of timber structures'
- EN 1996, 'Eurocode 6: Design of masonry structures'
- EN 1997, 'Eurocode 7: Geotechnical design'
- EN 1998, 'Eurocode 8: Design of structures for earthquake resistance'
- EN 1999, 'Eurocode 9: Design of aluminium structures'.

Status and field of application of Eurocodes

Generally, the Eurocodes provide structural design rules that may be applied to complete structures and structural components and other products. Rules are provided for common forms of construction, and it is recommended that specialist advice is sought when considering unusual structures. More specifically, the Eurocodes serve as reference documents that are recognised by the EU member states for the following purposes:

- as a means to prove compliance with the essential requirements of Council Directive 89/106/EEC

- as a basis for specifying contracts for construction or related works
- as a framework for developing harmonised technical specifications for construction products.

National standards implementing Eurocodes

The national standards implementing Eurocodes (e.g. BS-EN 1999-1-1: 2007 + A2) must comprise the full, unaltered text of that Eurocode and its annexes. Generally, there will be a national title page and a national foreword.

Of significant importance is the National Annex, which may be published as a separate document. The National Annex gives country-specific information on those parameters left open to national choice (e.g. values of partial safety factors). The National Annex also gives country-specific decisions on the status of informative annexes in the Eurocode – whether they become normative, remain informative and can be used, or whether they are not recommended for use in the country.

National choice is allowed in the clauses of EN 1999-1-1 listed in Table 1.

The National Annex may also reference non-contradictory complementary information. In the UK, PD 6702-1:2009 gives recommendations for the design of aluminium structures to BS EN 1999.

Links between Eurocodes and product-harmonised technical specifications (ENs and ETAs)

The clear need for consistency between the harmonised technical specifications for construction products and the technical notes for works is highlighted. Of particular note is that information

Table 1. Clauses in EN 1991-1-1 for which national choice is permitted

Subclause	Nationally Determined Parameter
1.1.2(1)	Minimum material thicknesses
2.1.2(3)	Options allowed by EN 1090 to suit reliability level required
2.3.1(1)	Actions for particular regional or climatic or accidental situations
3.2.1(1)	Use of aluminium alloys and tempers not listed in clause 3.2.1
3.2.2(1)	Rules for application of electrically welded tubes produced to EN 1592-1 to 4
3.2.2(2)	Characteristic strength at service temperatures between 80°C and 100°C
3.2.3.1(1)	Quality requirements for castings
3.3.2.1(3)	Provisions for the use of aluminium bolts and solid rivets
3.3.2.2(1)	Rules for preloaded bolts other than classes 8.8 and 10.9
5.2.1(3)	Global mode elastic instability criterion
5.3.2(3)	Design vales of initial bow imperfections
5.3.4(3)	Initial imperfection factor to be used for second order analysis taking account of lateral torsional buckling
6.1.3(1)	ULS partial safety factors
6.2.1(5)	Critical point yield criterion for the resistance of cross-sections
7.1(4)	Plastic redistribution of moments and force at serviceability limit state
7.2.1(1)	Building vertical deflection limits
7.2.2(1)	Building horizontal deflection limits
7.2.3(1)	Building dynamic effects limits
8.1.1(2)	Partial safety factors γ_M for joints
8.9(3)	Other joining methods
A.2	Rules for the application of consequence classes and reliability classes
C.3.4.1(2)	Partial safety factors γ_M for castings
C.3.4.1(3)	Partial safety factors γ_M for bearing resistance in castings with bolts and rivets
C.3.4.1(4)	Partial safety factors γ_M for resistance in castings with pin connections
K.1(1)	Specific flange geometry where shear lag effects can be ignored at ULS
K.3(1)	Methods for determining shear lag effects at ULS

accompanying the CE marking of construction products that refer to Eurocodes **must** detail which Nationally Determined Parameters have been taken into account.

Additional Information specific to EN 1999-1-1

As with the Eurocodes for other structural materials, Eurocode 9 is to be used in conjunction with EN 1990 and EN 1991 for basic principles, actions (loads) and combinations of actions.

EN1991-1-1 is the first of five parts of EN 1999. It gives the general design rules for most types of structure subject to predominately static actions. Other parts of Eurocode 9 deal with structural fire design, structures susceptible to fatigue, cold-formed sheeting and shell structures.

Designers' Guide to Eurocode 9: Design of Aluminium Structures
ISBN 978-0-7277-5737-1

ICE Publishing: All rights reserved
http://dx.doi.org/10.1680/das.57371.005

Chapter 1
General

Readers will note the similarities between Eurocode 9 and the other Eurocodes, and in particular many of the general clauses in this section are almost identical to those in Eurocode 3.

1.1. Scope

EN 1999 applies to the design of buildings and civil and structural engineering works in aluminium. It has to be used in conjunction with EN 1990 for the basis of design, and with EN 1991 for applied actions and combination of actions.

Comprehensive design rules are given for structures using wrought aluminium alloys with welded, bolted or riveted connections. Limited guidance is given for cast aluminium alloys and for adhesive-bonded connections.

The design rules cover a wide range of applications, and (unlike Eurocode 3) there are not separate parts for bridges, towers, crane-supporting structures, etc.

EN 1999 has five parts:

- EN 1999-1-1: 'Design of Aluminium Structures: General structural rules'
- EN 1990-1-2: 'Design of Aluminium Structures: Structural fire design'
- EN 1999-1-3: 'Design of Aluminium Structures: Structures susceptible to fatigue'
- EN 1999-1-4: 'Design of Aluminium Structures: Cold-formed structural sheeting'
- EN 1999-1-5: 'Design of Aluminium Structures: Shell structures'.

Part 1-1 has eight sections:

- Section 1: 'General'
- Section 2: 'Basis of design'
- Section 3: 'Materials'
- Section 4: 'Durability'
- Section 5: 'Structural analysis'
- Section 6: 'Ultimate limit states for members'
- Section 7: 'Serviceability limit states'
- Section 8: 'Design of joints'.

In addition, there are 13 annexes, all of which are informative except for Annex B. (Note that Annex A was originally normative, but was extended significantly in Amendment A1 to EN 1999-1-1, at which time it became informative.) The annexes cover the following:

- Annex A: 'Reliability differentiation'
- Annex B: 'Equivalent T-stub in tension'
- Annex C: 'Materials selection'
- Annex D: 'Corrosion and surface protection'
- Annex E: 'Analytical models for stress strain relationship'
- Annex F: 'Behaviour of cross section beyond elastic limit'
- Annex G: 'Rotation capacity'
- Annex H: 'Plastic hinge method for continuous beams'
- Annex I: 'Lateral torsional buckling of beams and torsional or flexural-torsional buckling of compression members'

- Annex J: 'Properties of cross sections'
- Annex K: 'Shear lag effects in member design'
- Annex L: 'Classification of connections'
- Annex M: 'Adhesive bonded connections'.

1.2. Normative references

A very large list of normative references is given in EN 1999-1-1, covering execution of structures, structural design, aluminium alloys, fasteners, welding and adhesives.

It should be noted that the majority of the normative references are dated, such that subsequent amendments or revisions to those references only apply to EN 1999-1-1 if incorporated in an amendment or revision to EN 1999-1-1.

1.3. Assumptions

The general assumptions given in EN 1990 apply to EN 1999, and cover the manner in which the structure is designed, constructed and maintained. They include the need for appropriate qualifications and skills of personnel and procedures for checking at all stages of design and execution.

EN 1999-1-1 specifically requires execution to be in accordance with EN 1090-3.

1.4. Distinction between principles and application rules

EN 1990 explicitly distinguishes between 'principles' and 'application rules'. Clause numbers that are followed by the letter 'P' are principles. In general, this notation is used in EN 1999-1-1 in clauses that invoke a high-level principle.

1.5. Terms and definitions

Clause 1.5

Terms and definitions are predominately covered in EN 1990. *Clause 1.5* of EN 1999-1-1 lists further definitions that are used. Some of these are also used in Eurocode 3, and, where possible, a consistent definition is given.

1.6. Symbols

Clause 1.6
Clause 6.7
Clause 6.8

Clause 1.6 lists the symbols used in the standard, ordered by the section in which they appear. Note that some symbols have different meanings in different sections (e.g. b_1 in *clause 6.7* is a distance from a stiffener, whereas b_1 in *clause 6.8* is a flange width). While the meanings are obvious when carrying out manual calculations, care should be taken in any highly computerised analysis.

Where possible, the symbols are chosen to be consistent with other Eurocodes.

Figure 1.1. Convention for member axes. (Reproduced from EN 1999-1-1)

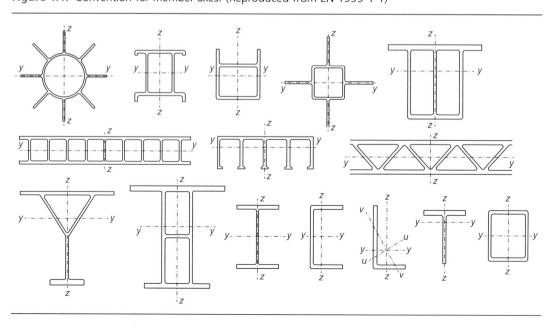

1.7.   Conventions for member axes

The conventions for member axes are the same as used in other Eurocodes: see Figure 1.1, reproduced from EN 1999-1-1.

Note that the design rules relate to principal axis properties, which for unsymmetrical sections differs from the x–x and y–y axes.

1.8.   Specification for execution of the work

Execution shall be carried out in accordance with EN 1090-3, and it is necessary to specify all of the information required to do so.

Annex A of EN 1090-3 lists required information, options to be specified and requirements related to execution class. In the UK, some guidance is given in PD 6705-3, 'Structural use of steel and aluminium – Part 3: Recommendations for the execution of aluminium structures to BS EN 1090-3'.

Designers' Guide to Eurocode 9: Design of Aluminium Structures
ISBN 978-0-7277-5737-1

Chapter 2
Basis of design

Readers will again note the similarities between Eurocode 9 and the other Eurocodes, and, in particular, many of the clauses in this section are almost identical to those in Eurocode 3.

2.1. Requirements
The basic requirement is that design is to be in accordance with the general rules of Eurocode 0, using the actions derived from Eurocode 1 and resistances from Eurocode 9. The standard also gives requirements for serviceability and durability.

Reliability management should follow the principles given in Eurocode 0, with execution in accordance with EN 1090-3. These require the designer to assess the consequences of failure and to choose relevant criteria for checking, execution, inspection and testing. Guidance on the relevant criteria is given in informative Annex A – see further detail in Chapter 9 of this guide. **Note:** Annex A is not recommended for use in the UK.

In addition to providing sufficient strength, the design should take into account durability, corrosion, fatigue, fire resistance and any applicable accidental actions. It should also ensure that allowance is made for all necessary inspection and maintenance.

2.2. Principles of limit state design
Eurocode 9 Part 1-1 gives resistances of members and cross-sections based on models of recognised experimental evidence for predominantly static loads. These resistances meet the ultimate limit states defined in Eurocode 0, and can therefore be used, subject to the conditions for materials given in Chapter 3 of this guide being met and execution being carried out in accordance with EN 1090-3.

2.3. Basic variables
Actions are to be taken from Eurocode 1 using the combinations and partial factors given in Annex A to Eurocode 0.

In addition, any actions during erection should be considered (Eurocode 1 Part 1-6), and the effects of settlements allowed for.

The effects of uneven settlements, imposed deformations and also any prestressing should be treated as permanent actions.

Fatigue loading should be derived using Eurocode 1 or using the rules given in Eurocode 9 Part 1-3. Note that the simplified approaches using damage-equivalent factors for fatigue loading given in parts of EN 1991 (e.g. for cranes) are not valid for aluminium as they are based on steel fatigue performance using an S–N slope of $m = 3$ for normal stress and $m = 5$ for shear stress.

2.4. Verification by the partial factor method
Material properties are given in Eurocode 9 for the range of permitted materials – see Chapter 3 of this guide.

Design resistances are based on γ_M, the partial factor for material properties that allows for model uncertainties and normal dimensional variations. The permitted tolerances and imperfections given in EN 1090-3 and referenced product standards are therefore taken into account. However, it is necessary to take account of deviations in geometric data where these are significant (e.g. as a result of non-linear behaviour, or the cumulative effects of multiple geometric deviations).

2.5. Design assisted by testing

Design may incorporate the results of testing, provided that the design resistances are determined in accordance with Annex D of EN 1990.

Designers' Guide to Eurocode 9: Design of Aluminium Structures
ISBN 978-0-7277-5737-1

ICE Publishing: All rights reserved
http://dx.doi.org/10.1680/das.57371.011

Chapter 3
Materials

This chapter concerns the guidance given in EN 1999-1-1 for materials as covered in *Section 3* of the code. The following clauses are addressed:

- General   *Clause 3.1*
- Structural aluminium   *Clause 3.2*
- Connecting devices   *Clause 3.3*

3.1. General

Material properties to be used in the design expressions throughout the code are based on the minimum values given in the relevant product standard. The properties are given in *Section 3* of EN 1999-1-1, and are specified as characteristic values.

3.2. Structural aluminium

There are a very large number of different aluminium alloys that can be obtained. Eurocode 9 lists the most commonly used alloys, and their properties, in the forms and tempers given in *Tables 3.1, 3.2* and *3.3*.

It is also possible to agree specific properties with the manufacturer for the production of material for a specific contract, although care should be exercised in using any other alloy or non-standard material property. Any item that is placed on the market as designed in accordance with Eurocode 9 should only use the alloys and properties listed in *Section 3*.

Guidance on the choice of a suitable alloy for any particular application is given in *Annex C*.

Wrought aluminium alloys for structures are listed in *Table 3.1a* for the following products:

- sheet (SH), strip (ST) and plate (PL)   EN 485
- extruded tubes (ET), hollow profiles (EP/H), open profiles (EP/O), rods and bars
  (ER/B)   EN 755
- drawn tubes (DT)   EN 754
- forgings (FO)   EN 586

EN 1999-1-1 has limited applicability to castings. However, a number of cast aluminium alloys are listed in *Table 3.1b*. *Annex C* gives further information for the design of structures using cast aluminium alloys.

The alloys listed in *Tables 3.1a* and *3.1b* are categorised into the three durability ratings A, B and C, in descending order of durability. These ratings are used to determine the need for any protection required in different environments – see *Annex D* (see Section 9.4 of this guide).

Characteristic values of the 0.2% proof strength f_o and the ultimate tensile strength f_u for wrought aluminium alloys for a range of tempers and thicknesses are given in:

- *Table 3.2a* for sheet, strip and plate products
- *Table 3.2b* for extruded rod/bar, extruded tube, extruded profiles and drawn tube
- *Table 3.2c* for forgings.

Note that the strengths vary with product form and with thickness. Therefore, if it is not certain at the design stage whether a member will be fabricated, for example, from a plate or an extruded flat, then the lower-strength property should be used.

Clause 6.1.6

Characteristic values for strength in the heat-affected zone (HAZ) (0.2% proof strength $f_{o,haz}$, and ultimate tensile strength $f_{u,haz}$) are given in the tables together with reduction factors for HAZ (see *clause 6.1.6*). Note that the HAZ values and reduction factors are only valid for MIG welding of elements up to 15 mm thick. For TIG welding and for greater thickness, it is necessary to apply a larger reduction factor (see the footnotes to *Tables 3.2a, 3.2b* and *3.2c*).

Clause 6.1.4
Clause 6.3.1

The buckling class (used in *clauses 6.1.4* and *6.3.1*) and the exponent in the Ramberg–Osgood expression for plastic resistance are also listed in the tables.

Clause 3.2.2

The properties are suitable for use in structures that will experience service temperatures up to 80°C. *Clause 3.2.2* gives a formula for calculating a reduction factor if the temperature will be between 80°C and 100°C. For temperatures over 100°C a reduction of the elastic modulus and additionally a time dependant, non-recoverable reduction of strength should be considered. For guidance on these reductions and for structural fire design, see EN 1999-1-2.

Clause 3.2.3

Characteristic values for 0.2% proof strength f_{oc} and the ultimate tensile strength f_{uc} of cast aluminium alloys are given in *clause 3.2.3* (Table 3.3). Note that these values differ from the required strength of test specimens as given by EN 1706.

Table 3.1 gives values for four frequently used aluminium alloys as examples:

- EN AW-6063 is very suited for decorative anodising, and is used if strength is not of paramount importance.
- EN AW-6005A is a medium-strength alloy able to be extruded into very complex shapes.
- EN AW-6082 is widely used for welded and non-welded applications where high strength, good corrosion resistance and good machining properties are required.
- EN AW-5083 is a strong alloy that has excellent corrosion resistance and good strength in the HAZ at welds. It is used in marine environments and for structures welded from plates.

In *Table 3.2b* (and Table 3.1) some values are quoted in bold. For these values, greater thicknesses and/or higher mechanical properties may be permitted in some forms according to the applicable EN standard.

Clause 3.2.5

The material constants to be adopted in calculations for the aluminium alloys covered by the standard are given in *clause 3.2.5* as follows:

- modulus of elasticity $E = 70\,000$ N/mm^2
- shear modulus $G = 27\,000$ N/mm^2
- Poisson's ratio $\nu = 0.3$
- coefficient of linear thermal expansion $\alpha = 23 \times 10^{-6}/°C$
- unit mass $\rho = 2700$ kg/m^3.

3.3. Connecting devices

Clause 3.3

Clause 3.3 gives requirements for connecting devices, including bolts, friction grip fasteners, solid rivets, special fasteners, welds and adhesives. References are given to EN and ISO standards or, for solid rivets, to recommendations in *Annex C*.

Requirements for self-tapping and self-drilling screws and blind rivets used for thin-walled structures are given in EN 1999-1-4.

Table 3.4 gives values of 0.2% proof strength $f_{o,haz}$ and ultimate tensile strength $f_{u,haz}$ for aluminium alloy, steel and stainless steel bolts and solid rivets for use in calculating the design resistance in *Section 8*.

Clause 3.3.4

Some guidance on the selection of filler metal for welds is given in *clause 3.3.4*. EN 1011-4 gives more comprehensive information.

Table 3.1. Characteristic values of strength, minimum elongation, reduction factors in HAZ, buckling class BC and exponent n_p for four examples of wrought aluminium alloys:
- extruded profiles (EP, EP/O, EP/H), extruded tube (ET), extruded rod/bar (ER/B) and drawn tube (DT) (data from Table 3.2a in EN 1999-1-1)
- sheet (SH), strip (ST) and plate (PL) (data from Table 3.2b in EN 1999-1-1)

Alloy	Product form	Temper	Thickness t: mm	f_o: N/mm²	f_u: N/mm²	A: %	$f_{o,haz}$: N/mm²	$f_{u,haz}$: N/mm²	HAZ factor		BC	n_p
									$\rho_{o,haz}$	$\rho_{u,haz}$		
EN AW-6063	EP, ET, ER/B	T5	$t \le 3$	130	175	8	60	100	0.46	0.57	B	16
	EP		$3 < t \le 25$	110	160	7	60	100	0.55	0.63	B	13
	EP, ET, ER/B	T6	$t \le 25$	160	195	8	65	110	0.41	0.56	A	24
	DT		$t \le 20$	190	220	10	65	110	0.34	0.50	A	31
EN AW-6005A	EP/O, ER/B	T6	$t \le 5$	225	270	8	115	165	0.51	0.61	A	25
			$5 < t \le 10$	215	260	8	115	165	0.53	0.63	A	24
			$10 < t \le 25$	200	250	8	115	165	0.58	0.66	A	20
	EP/H, ET	T6	$t \le 5$	215	255	8	115	165	0.53	0.65	A	26
			$5 < t \le 10$	200	250	8	115	165	0.58	0.66	A	20
EN AW-6082	EP, ET, ER/B	T4	$t \le 25$	110	205	14	100	160	0.91	0.78	B	8
	EP/O, EP/H	T5	$t \le 5$	230	270	8	125	185	0.54	0.69	B	28
	EP/O, EP/H, ET	T6	$t \le 5$	250	290	8	125	185	0.50	0.64	A	32
			$5 < t \le 15$	260	310	10	125	185	0.48	0.60	A	25
	ER/B	T6	$t \le 20$	250	295	8	125	185	0.50	0.63	A	27
			$20 < t \le 150$	260	310	8	125	185	0.48	0.60	A	25
EN AW-5083	SH/ST/PL	O/H111	≤50	125	275	11	125	275	1	1	B	6
		H12	≤40	250	305	3	155	275	0.62	0.9	B	22
		H14	≤25	280	340	2	155	275	0.55	0.81	A	22

Designers' Guide to Eurocode 9: Design of Aluminium Structures
ISBN 978-0-7277-5737-1

ICE Publishing: All rights reserved
http://dx.doi.org/10.1680/das.57371.015

Chapter 4
Durability

The basic requirements for durability are given in Eurocode 0, which states that:

> The structure shall be designed such that deterioration over its design working life does not impair the performance of the structure below that intended, having due regard to its environment and the anticipated level of maintenance.

Structures made of the aluminium alloys listed in *Section 3* generally do not need any protective treatment to maintain structural integrity for typical lives of buildings and civil engineering structures in normal atmospheric conditions.

Areas or environments that give conditions where protective treatment is likely to be required include:

- structures to be used in severe industrial or polluted marine environments
- structures that will be subject to immersion in water
- parts of structures in contact with concrete or plaster
- parts of structures in contact with other metals
- parts of structures in contact with soil
- parts of structures in contact with certain species of timber.

Guidance on the durability of different alloys and when protective treatment is recommended is given in *Annexes C* and *D* (see Chapter 9 of this guide).

While *Section 2* has a general requirement that allowance is made for all necessary inspection and maintenance, this section specially notes that components should be designed such that inspection, maintenance and repair can be carried out satisfactorily during the design life of the structure if they are susceptible to corrosion, mechanical wear or fatigue.

Requirements for the execution of protective treatment are given in EN 1090-3.

Recommendations for the choice of mechanical fasteners for structural sheeting to avoid corrosion are given in *Annex B* of EN 1999-1-4 (see Section 10.10 of this guide).

Designers' Guide to Eurocode 9: Design of Aluminium Structures
ISBN 978-0-7277-5737-1

ICE Publishing: All rights reserved
http://dx.doi.org/10.1680/das.57371.017

Chapter 5
Structural analysis

Eurocode 9 provides a level of detail that is not given in previous aluminium design codes used in the UK with regard to specifying aspects to be taken into account in the structural analysis that is used to determine the forces and moments in members and joints. The provisions are almost identical to those in Eurocode 3, and so will be familiar to designers acquainted with current steel design codes. In particular, there are requirements that cover the choice of elastic or plastic global analysis, joint rigidity and second-order effects.

The following clauses are addressed in this chapter:

- Structural modelling   *Clause 5.1*
- Global analysis   *Clause 5.2*
- Imperfections   *Clause 5.3*
- Methods of analysis   *Clause 5.4*

5.1. Structural modelling

Clause 5.1 requires that the modelling should be accurate and appropriate for the limit state under consideration, which effectively dictates that elastic analysis should be used for the consideration of serviceability criteria, and implies that second-order effects should be considered when deflections are large.

Clause 5.1

The effects of joint rigidity may need to be taken into account in the analysis, depending on whether the joints are simple joints that cannot transmit bending moments, continuous joints that can give full strength and stiffness, or semi-continuous joints that give some stiffness but are insufficient to be considered continuous. Further guidance is given in *Annex L* and in Section 9.12 of this guide.

Where appropriate, the deformation of supports should be allowed for. This may be deformation of a structure formed of other materials, or may be in relation to interaction with the ground. If the latter, Eurocode 7 (EN 1997) should be referred to for guidance on soil–structure interaction.

5.2. Global analysis

One of the first decisions is whether second-order analysis is necessary. Often it will be obvious: for example, for a stiff, fully braced structure, first-order analysis will generally be sufficient, whereas structures that may deflect or sway by significant amounts will generally require a second-order analysis. If there is doubt, then it will be necessary to use computer software to determine the elastic critical load for the structure, and then to check this against the limit given in *Equation 5.1*:

$$\alpha_{cr} = \frac{F_{cr}}{F_{Ed}} \leq 10 \qquad (5.1)$$

where:

α_{cr} is the factor by which the design loading would have to be increased to cause elastic instability in a global mode

F_{Ed} is the design loading on the structure

F_{cr} is the elastic critical buckling load for global instability mode based on initial elastic stiffness.

Clause 5.2.2

The model used in the global analysis should make suitable allowances for the flexibility arising from shear lag, from local buckling and joint flexibility as appropriate. If it is necessary to apply a second-order analysis, the standard gives three alternative methods in *clause 5.2.2*. These are:

Clause 6.3

- To account for all global and local geometric and material imperfections in a second-order global analysis. Such an analysis involves complex computer software that will account for global frame instability as well as member buckling. If this approach is taken, then individual member buckling checks in accordance with *clause 6.3* will not be necessary.

Clause 6.3

- To account for global second order effects such as frame imperfections and sway in a second-order global analysis, and to deal separately with member buckling in accordance with *clause 6.3*.

Clause 5.3
Clause 6.3

- To account for global second-order effects by enhancing the moments and forces calculated using a linear analysis by applying equivalent forces and/or equivalent members in accordance with *clause 5.3*, and to deal separately with member buckling in accordance with *clause 6.3*.

Clause 5.2.2
Clause 6.3
Clause 5.3

Note that *clause 5.2.2* incorrectly refers to *clause 6.3* rather than to *clause 5.3* for equivalent members and the equivalent column method.

5.3. Imperfections

Clause 5.3.2(1)
Clause 5.3.2(11)

The assumed shape of imperfections may be derived from an analysis of the elastic buckling mode of the members and structure under consideration (*clause 5.3.2(1)* and *clause 5.3.2(11)*).

*Clause 5.3.2(3)–
5.3.2(6)*

Alternatively, details of the geometric allowances for imperfections for sway of frames and the bow in members that are liable to buckle (referred to as equivalent imperfections) that are to be incorporated in the analysis can be taken from the rules given in *clauses 5.3.2(3)* to *5.3.2(6)*.

*Clause 5.3.2(7)–
5.3.2(10)*

A further alternative is given whereby equivalent horizontal forces are applied in lieu of geometric allowances (*clauses 5.3.2(7)* to *5.3.2(10)*): see Figures 5.1 and 5.2. ϕ is a sway imperfection obtained from the expression

$$\phi = \phi_0 \alpha_h \alpha_m \qquad (5.2)$$

where:

ϕ_0 is the basic value, $\phi_0 = 1/200$

Figure 5.1. Configuration of sway imperfections ϕ for horizontal forces on floor diaphragms. (a) Two or more storeys. (b) Single storey. (Reproduced from EN 1999-1-1 (Figure 5.2), with permission from BSI)

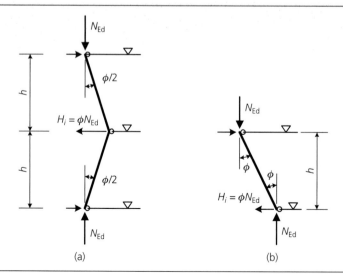

(a)   (b)

Figure 5.2. Replacement of initial imperfections by equivalent horizontal forces. (a) Initial sway imperfections. (b) Initial bow imperfections. (Reproduced from EN 1999-1-1 (Figure 5.3), with permission from BSI)

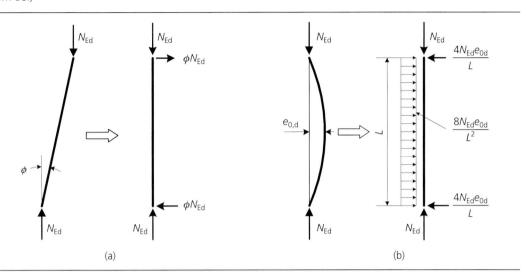

α_h is the reduction factor for height h applicable to columns,

$$\alpha_h = \frac{2}{\sqrt{h}} \quad \text{but} \quad \frac{2}{3} \leq \alpha_h \leq 1.0$$

h is the height of the structure in metres

α_m is the reduction factor for the number of columns in a row,

$$\alpha_m = \sqrt{0.5 \left(1 + \frac{1}{m}\right)}$$

m is the number of columns in a row including only those columns that carry a vertical load N_{Ed} not less than 50% of the average value of the column in the vertical plane considered.

The initial bow imperfection e_0 is determined from the ratio to member length L given in Table 5.1.

In similar manner, *clause 5.3.3* gives details of geometric allowances or equivalent forces that can be used for bracing systems used to give restraint to beams or compression members.

Clause 5.3.3

5.4.   Methods of analysis

Member forces and moments can be determined using elastic analysis in all cases, and will generally give an acceptable solution whereby superposition of results from various load cases can be readily applied.

Alternatively, a plastic analysis can be used if the following conditions are met:

■ there is sufficient rotational capacity at plastic hinge locations
■ there is no buckling of members within the structure
■ there are no welds at potential hinge locations in areas of tensile stress.

Table 5.1. Design values of initial bow imperfection e_0/L

Buckling class according to *Table 3.2*	Elastic analysis, e_0/L	Plastic analysis, e_0/L
A	1/300	1/250
B	1/200	1/150

Guidance that is useful in any consideration of plastic analysis is given in *Annex G* (see Section 9.7 of this guide) regarding rotation capacity, *Annex H* (see Section 9.8 of this guide) regarding plastic hinges in continuous beams and *Annex L* (see Section 9.12 of this guide) regarding classification of joints.

Designers' Guide to Eurocode 9: Design of Aluminium Structures
ISBN 978-0-7277-5737-1

ICE Publishing: All rights reserved
http://dx.doi.org/10.1680/das.57371.021

Chapter 6
Ultimate limit states

This chapter concerns the subject of cross-section, member and plate design at the ultimate limit state. The material in this chapter is covered in *Section 6* of EN 1999-1-1, and the following clauses are addressed:

<table>
<tr><td>■ Basis</td><td>*Clause 6.1*</td></tr>
<tr><td>■ Resistance of cross-section</td><td>*Clause 6.2*</td></tr>
<tr><td>■ Buckling resistance of members</td><td>*Clause 6.3*</td></tr>
<tr><td>■ Uniform built-up members</td><td>*Clause 6.4*</td></tr>
<tr><td>■ Unstiffened plates under in-plane loading</td><td>*Clause 6.5*</td></tr>
<tr><td>■ Stiffened plates under in-plane loading</td><td>*Clause 6.6*</td></tr>
<tr><td>■ Plate girders</td><td>*Clause 6.7*</td></tr>
<tr><td>■ Members with corrugated web</td><td>*Clause 6.8*</td></tr>
</table>

EN 1999-1-1 is a comprehensive and in most parts a stand-alone document. This chapter is first of all focused on the cross-section and member design, including plate girders. Parts of it are similar to BS 8118 (BSI, 1991).

6.1. Basis
6.1.1 General
Aluminium structures and components shall be proportioned so that the basic design requirements for the ultimate limit state given in *Section 2* are satisfied. The design recommendations are for structures subjected to normal atmospheric conditions.

6.1.2 Characteristic value of strength
Resistance calculations for members are made using characteristic values of strength, as follows:

■ f_o is the characteristic value of the strength for bending and overall yielding in tension and compression
■ f_u is the characteristic value of the strength for the local resistance of a net section in tension or compression.

The characteristic values of the 0.2% proof strength f_o and the ultimate tensile strength f_u for wrought aluminium alloys are given in the material standards. These are given in *clause 3.2.2* for selected structural aluminium alloys.

Clause 3.2.2

6.1.3 Partial safety factors
In the structural Eurocodes, partial factors γ_M are applied to different components in various situations to reduce their resistances from characteristic values to design values (or, in practice, to ensure that the required level of safety is achieved). The uncertainties (material, geometry, modelling, etc.) associated with the prediction of resistance for a given case, as well as the chosen resistance model, dictate the value of γ_M that is to be applied. Partial factors are discussed in Section 2.4 of this guide, and in more detail in EN 1991 and elsewhere. γ_M factors assigned to particular resistances in EN 1999-1-1 according to *clause 6.1.3* are given in Table 6.1 as well as recommended numerical values. However, for structures to be constructed in particular countries in Europe, reference should be made to the National Annexes, which might prescribe modified values.

Clause 6.1.3

Table 6.1. Partial safety factors for ultimate limit states recommended in Eurocode 9 (data from EN 1999-1-1, *Table 6.1*)

Resistance of cross-sections (whatever the class)	$\gamma_{M1} = 1.10$
Resistance of members to instability assessed by member checks	$\gamma_{M1} = 1.10$
Resistance of cross-sections in tension to fracture	$\gamma_{M2} = 1.25$
Resistance of joints (bolt and rivet connections, plates in bearing)	$\gamma_{M2} = 1.25$
Resistance of other connections	See Chapter 8 of this guide

Note that there is no distinction between the partial factor for the resistance of cross-sections and for the instability of members in Eurocode 9, which means that there is a smooth transition between these two cases.

6.1.4 Classification of cross-sections

Basis

Clause 6.1.4
Clause 6.1.5
Clause 6.1.6

Clause 6.1.4
Clause 6.1.5

Determining the resistance of structural aluminium components requires the designer to consider first the cross-sectional behaviour and second the overall member behaviour. ***Clauses 6.1.4, 6.1.5*** and ***6.1.6*** cover the cross-sectional aspects of the design process. Whether in the elastic or the inelastic material range, cross-sectional resistance and rotation capacity are limited by the effect of local buckling. The code accounts for the effect of local buckling through cross-sectional classification, as described in ***clause 6.1.4***. Cross-sectional resistances may then be determined from ***clause 6.1.5***.

Clause 6.1.6

Clause 6.1.4.4(3)

In the design of welded structures using strain-hardened or artificially aged precipitation hardening alloys the reduction in strength properties that occurs in the vicinity of welds shall be allowed for (see ***clause 6.1.6***). The reduced strength in the heat-affected zone (HAZ) due to welds influences the cross-section classification and the determination of the resistance. Note that transverse welds can be ignored in determining the slenderness parameter, provided that there is lateral restraint at the weld location (***clause 6.1.4.4(3)***).

In Eurocode 9, cross-sections are placed into one of four behavioural classes, depending on the material proof strength, the width-to-thickness ratios of the individual compression parts (e.g. webs and flanges) within the cross-section, the presence of welds and the loading arrangement.

Definition of classes

Clause 6.1.4.2

Four classes of cross-sections are defined in Eurocode 9, as follows (***clause 6.1.4.2***):

- Class 1 cross-sections are those that can form a plastic hinge with the rotation capacity required for plastic analysis without reduction of the resistance.
- Class 2 cross-sections are those that can develop their plastic moment resistance, but have limited rotation capacity because of local buckling.
- Class 3 cross-sections are those in which the calculated stress in the extreme compression fibre of the aluminium member can reach its proof strength, but local buckling is liable to prevent development of the full plastic moment resistance.
- Class 4 cross-sections are those in which local buckling will occur before the attainment of proof stress in one or more parts of the cross-section.

The moment–rotation characteristics of the four classes are shown in Figure 6.1. Class 1 cross-sections are fully effective under pure compression and are capable of reaching, and even exceeding (see *Annex G*), the full plastic moment in bending, and may therefore be used in plastic design. Class 2 cross-sections have a somewhat lower deformation capacity, and are capable of reaching their full plastic moment in bending. Class 3 cross-sections are fully effective in pure compression, but local buckling prevents attainment of the full plastic moment in bending. Bending moment resistance lies between the plastic and the elastic moment, depending on the slenderness of the most slender part of the cross-section. For class 4 cross-sections, local buckling occurs in the elastic range. The effective cross-section is therefore defined based on the width-to-thickness ratios of individual cross-section parts. This effective cross-section is then used to determine the cross-sectional resistance. Unlike

Figure 6.1. Classification of cross-section according to Eurocode 9 and corresponding stress distribution

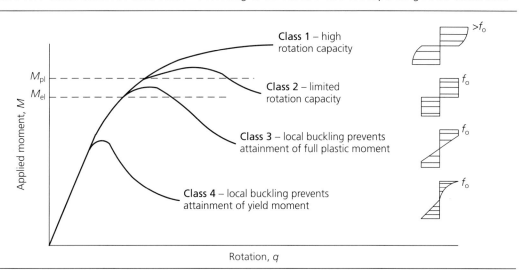

Eurocode 3 (steel), the effective thickness is used in Eurocode 9 instead of the effective width to build up the effective cross-section.

Assessment of individual parts

Each compressed or partially compressed cross-section part is assessed individually against the limiting width-to-thickness ratios for class 1, 2 and 3 elements defined in Table 6.4 (see *Table 6.2* in ***clause 6.1.4.4***). In the table, separate values are given for internal cross-section parts (defined as those supported along each edge by an adjoining flange or web) and for outstand cross-section parts where one edge of the part is supported by an adjoining flange or web and the other edge is free.

Clause 6.1.4.4

The limiting width-to-thickness ratios are modified with a factor ε that is dependent on the material proof strength defined as

$$\varepsilon = \sqrt{\frac{250\,\mathrm{MPa}}{f_\mathrm{o}}} \qquad\qquad\text{(D6.1)}$$

where f_o is the characteristic value of the proof strength as defined in *Tables 3.2a* and *3.2b* (see the examples in Table 3.1 of this guide). It may be of interest to notice that in Eurocode 3 the basic value in the expression of ε is 235 MPa, compared with 250 MPa in Eurocode 9.

The various compression parts in a cross-section (such as a web or a flange) can, in general, be in different classes. A cross-section is classified according to the highest (least favourable) class of its compression parts.

Three basic types of thin-walled parts are identified in the classification process according to ***clause 6.1.4.2***: flat outstand parts, flat internal parts and curved internal parts. These parts can be un-reinforced or reinforced by longitudinal stiffening ribs or edge lips or bulbs (see *Figure 6.1* in the code).

Clause 6.1.4.2

For outstand cross-section parts, b is the width of the flat part outside the fillet. For internal parts, b is the flat part between the fillets, except for rounded outside corners (see Figure 6.2).

The slenderness β for flat compression parts are given in Table 6.2, based on expressions in ***clause 6.1.4.3(1)***. In the same clause the parameter is also given for cross-section parts with reinforcement of a single-sided rib or lip of thickness equal to the thickness of the cross-section part (standard reinforcement), and methods on how to treat non-standard reinforcement and complex reinforcement are provided. Furthermore, the slenderness β for uniformly

Clause 6.1.4.3(1)

Figure 6.2 Definition of the width b (b_f for flanges and b_w for webs) for internal (I) and outstand (O) cross-section parts and corner details

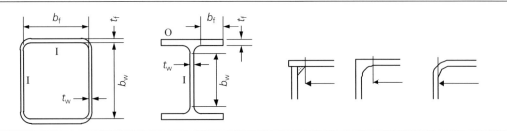

Table 6.2. Slenderness β for flat cross-section parts

Type of cross-section part	$\psi < -1$	$\psi = -1$	$-1 < \psi < 1$	$\psi = 1$
Internal cross-section part or outstand part with peak compression at root	$\dfrac{0.8}{(1-\psi)}\dfrac{b}{t}$	$0.4\dfrac{b}{t}$	$(0.7+0.3\psi)\dfrac{b}{t}$	$\dfrac{b}{t}$
Outstand part with peak compression at toe	$\dfrac{b}{t}$	$\dfrac{b}{t}$	$\dfrac{b}{t}$	$\dfrac{b}{t}$

Clause 6.1.4.3(4)
Clause 6.1.4.3(5)

compressed shallow curved unreinforced internal parts and thin-walled round tubes, whether in uniform compression or in bending, are given in *clauses 6.1.4.3(4)* and *6.1.4.3(5)* (see Table 6.3).

In Table 6.2, ψ is the ratio of the stresses at the edges of the plate under consideration related to the maximum compressive stress. In general, the neutral axis should be the elastic neutral axis, but in checking whether a section is class 1 or 2 it is permissible to use the plastic neutral axis. If the width of the part in compression is b_c, then the following formula may be used in classifying the cross-section part.

$$\psi = 1 - \frac{b}{b_c} \tag{D6.2}$$

The classification limits

Clause 6.1.4.4

Classification limits are given in Table 6.4 (*Table 6.2* in Eurocode 9 *clause 6.1.4.4*) for internal and outstand parts, Mazzolani *et al.* (1996). Values are dependent on the material buckling class A or B, according to *Table 3.2* in *clause 3.2.2* (see examples in Table 3.1 of this guide) and whether the member is longitudinally welded or not. In members with longitudinal welds, the classification is independent of the extent of the HAZ. Furthermore, a cross-section part may be considered as without welds if the welds are transverse to the member axis and located at a position of lateral restraint.

Clause 3.2.2

Table 6.3. Slenderness β for curved cross-section parts

Shallow curved unreinforced internal part	Thin-walled round tube
$\beta = \dfrac{b}{t}\dfrac{1}{\sqrt{1+0.006\dfrac{b^4}{R^2 t^2}}}$	$\beta = 3\sqrt{\dfrac{D}{t}}$

Table 6.4. Slenderness limits β_1/ε, β_2/ε and β_3/ε (data from EN 1999-1-1, *Table 6.2*)

Material classification according to Table 3.2	Internal part			Outstand part		
	β_1/ε	β_2/ε	β_3/ε	β_1/ε	β_2/ε	β_3/ε
Class A, without welds	11	16	22	3	4.5	6
Class A, with welds	9	13	18	2.5	4	5
Class B, without welds	13	16.5	18	3.5	4.5	5
Class B, with welds	10	13.5	15	3	3.5	4

$\varepsilon = \sqrt{250/f_o}$, f_o in N/mm^2.

The classification limits provided in Table 6.4 assume that the cross-section is stressed to yield, although, where this is not the case, **clause 6.1.4.4(4)** allows some modification when parts are less highly stressed. A modified expression $\varepsilon = \sqrt{(250/f_o)(z_1/z_2)}$ may be used to increase the limits. In this expression, z_1 is the distance from the elastic neutral axis of the effective section to the most severely stressed fibres (tension or compression), and z_2 is the distance from the elastic neutral axis of the effective section to the part under consideration. z_1 and z_2 should be evaluated on the effective section by means of an iterative procedure (minimum of two steps). The possibility of modification of the basic definition of ε given by Equation D6.1 (and thus the value of the classification limits) if the stress of the applied load is less than the proof stress f_o is not given in Eurocode 9 as it is in Eurocode 3 in certain cases.

Clause 6.1.4.4(4)

Overall cross-section classification

Once the classification of the individual parts of the cross-section is determined, the cross-section is classified according to the highest (least favourable) class of its compression parts. However, it should be noted that the classification of cross-sections for members in combined bending and axial compression is made for the loading components separately. (See the notes to **clause 6.3.3(4)**.) No classification is needed for the combined state of stress. This means that a cross-section can belong to different classes for axial force, major axis bending and minor axis bending. The combined state of stress is allowed for in the interaction expressions that should be used for all classes of cross-section.

Clause 6.3.3(4)

6.1.5   Local buckling resistance

Local buckling of class 4 members is generally allowed for by replacing the true section by an effective section. The effective section is obtained by employing a local buckling factor ρ_c to factor down the thickness. ρ_c is applied to any uniform-thickness class 4 part that is wholly or partly in compression. Parts that are not uniform in thickness require a special study.

The factor ρ_c is given by expression *6.11* or *6.12*, separately for different parts of the section, in terms of the ratio β/ε, where β is found in Table 6.2 or 6.3 (or in **clause 6.1.4.3(2)** or **6.1.4.3(3)** for stiffened cross-section parts), ε is defined in Equation D6.1, and the constants C_1 and C_2 in Table 6.5 (*Table 6.3* in **clause 6.1.5**).

Clause 6.1.4.3(2)
Clause 6.1.4.3(3)

Clause 6.1.5

$$\rho_c = 1.0 \qquad \text{if } \beta \leq \beta_3 \qquad\qquad (6.11)$$

$$\rho_c = \frac{C_1}{\beta/\varepsilon} - \frac{C_2}{(\beta/\varepsilon)^2} \qquad \text{if } \beta > \beta_3 \qquad\qquad (6.12)$$

Table 6.5. Constants C_1 and C_2 in expressions for ρ_c (data from EN 1999-1-1, *Table 6.3*)

Material classification according to Table 3.2	Internal part		Outstand part	
	C_1	C_2	C_1	C_2
Class A, without welds	32	220	10	24
Class A, with welds	29	198	9	20
Class B, without welds	29	198	9	20
Class B, with welds	25	150	8	16

Note that for flat outstand parts in unsymmetrical cross-sections (e.g. channels), ρ_c is given by the above expressions for flat outstand in symmetrical sections, but is not more than $120/(\beta/\varepsilon)^2$.

For reinforced cross-section parts, consider all possible modes of buckling and take the lower value of ρ_c. In the case of mode 1 buckling (distortional buckling, see *Figure 6.3;* of the code) the factor ρ_c should be applied to the area of the reinforcement as well as to the basic plate thickness. For reinforced outstand cross-section parts, use the curve for outstands, otherwise use the curve for an internal cross-section part.

The reduction factor ρ_c in sections required to carry biaxial bending or combined bending and axial load may have different values for the effective cross-sections for the separate loadings.

6.1.6   HAZ softening adjacent to welds

General

Welded structures using strain-hardened or artificially aged precipitation hardening alloys suffer from a reduction in strength properties in the vicinity of welds. Exceptions, where there is no weakening adjacent to welds, are alloys in the O condition or in the F condition, if the design strength is based on O condition properties. For design purposes, it is assumed that the strength properties are reduced by a constant level throughout the HAZ.

Even small welds to connect a small attachment to a main member may considerably reduce the resistance of the member due to the presence of a HAZ. In beam design it is often beneficial to locate welds and attachments in low-stress areas (i.e. near the neutral axis or away from regions of high bending moment).

For some heat-treatable alloys, it is possible to mitigate the effects of HAZ softening by means of artificial ageing applied after welding. However, no values of the mitigation are given in Eurocode 9.

Severity of softening

Clause 3.1.2

The characteristic value of the 0.2% proof strengths $f_{o,haz}$ and the ultimate strength $f_{u,haz}$ in the HAZ are listed in *Tables 3.2a, 3.2b* and *3.2c* in **clause 3.1.2** (examples are given in Table 3.1 of this guide), which also gives the reduction factors

$$\rho_{o,haz} = \frac{f_{o,haz}}{f_o} \tag{6.13}$$

$$\rho_{u,haz} = \frac{f_{u,haz}}{f_u} \tag{6.14}$$

The reduction affects the 0.2% proof strength of the material more severely than the ultimate tensile strength. The affected region extends immediately around the weld, beyond which the strength properties rapidly recover to their full non-welded values.

Material in the HAZ recovers some strength after welding due to a natural ageing process, and the values of $f_{o,haz}$ and $f_{a,haz}$ in *Tables 3.2a, 3.2b* and *3.2c* are only valid from at least 3 days after welding for 6xxx series alloys and 30 days after welding for 7xxx series alloys, providing the material has been held at a temperature not less than 10°C. If the material is held at a temperature below 10°C after welding, the recovery time will be prolonged.

The severity of softening can be taken into account by the characteristic value of strength $f_{o,haz}$ and $f_{u,haz}$ in the HAZ metal using the full cross-section. This method is used in the design of joints (see Chapter 8). For member design, Eurocode 9 accounts for the HAZ by reducing the assumed cross-sectional area over which the stresses acts with the factors $\rho_{o,haz}$ or $\rho_{u,haz}$ over the width of the HAZ (b_{haz}). This is especially convenient, as local buckling is allowed for by an effective thickness ($t_{eff} = \rho_c t$) as well. (See later in this guide.)

Extent of the HAZ

Clause 6.1.6.3

The HAZ is assumed to extend a distance b_{haz} in any direction from a weld, measured as follows (see the example in Figure 6.3 – *Figure 6.6;* in **clause 6.1.6.3**):

Figure 6.3. The extent of HAZs. (Reproduced from EN 1999-1-1 (*Figure 6.6*), with permission from BSI)

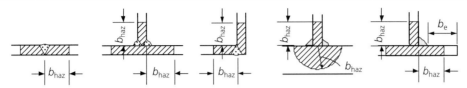

If the distance b_e is less than $3b_{haz}$, assume that the HAZ extends to the full width of the outstand

- transversely from the centreline of an in-line butt weld
- transversely from the point of intersection of the welded surfaces at fillet welds
- transversely from the point of intersection of the welded surfaces at butt welds used in corner, tee or cruciform joints
- in any radial direction from the end of a weld.

The HAZ boundaries should generally be taken as straight lines normal to the metal surface, particularly if welding thin material. However, if surface welding is applied to thick material, it is permissible to assume a curved boundary of radius b_{haz}, as shown in Figure 6.3.

For a MIG weld laid on unheated material, and with interpass cooling to 60°C or less when multi-pass welds are laid, values of b_{haz} are given in Table 6.6, based on *clause 6.1.6.3*. For a thickness >1 mm there may be a temperature effect, because interpass cooling may exceed 60°C unless there is strict quality control. This will increase the width of the HAZ.

Clause 6.1.6.3

For a TIG weld the extent of the HAZ is greater because the heat input is higher than for a MIG weld, and has a value of b_{haz} given in Table 6.6.

The values in the table apply to in-line butt welds (two valid heat paths) or to fillet welds at T junctions (three valid heat paths) in 6xxx and 7xxx series alloys, and in 3xxx and 5xxx series alloys in the work-hardened condition.

If two or more welds are close to each other, their HAZ boundaries overlap. A single HAZ then exists for the entire group of welds. If a weld is located too close to the free edge of an outstand, the dispersal of heat is less effective. This applies if the distance from the edge of the weld to the free edge is less than $3b_{haz}$. In these circumstances, assume that the entire width of the outstand is subject to the factor $\rho_{o,haz}$.

Other factors that affect the value of b_{haz} for which information is given in *clause 6.1.6.3(8)* are:

Clause 6.1.6.3(8)

- the influence of temperatures above 60°C
- variations in thickness
- variations in the number of heat paths.

6.2.   Resistance of cross-sections
6.2.1   General
Prior to determining the resistance of a cross-section, the cross-section should be classified in accordance with *clause 6.1.4*. Cross-section classification is described in detail in Section 6.1.4

Clause 6.1.4

Table 6.6. Extent of b_{haz} for MIG and TIG welds

Thickness	MIG	TIG
$0 < t \le 6$ mm	20 mm	30 mm
$6 < t \le 12$ mm	30 mm	
$12 < t \le 25$ mm	35 mm	
$t > 25$ mm	40 mm	

Clause 6.2

of this guide. *Clause 6.2* covers the resistance of cross-sections, including the resistance to tensile fracture at net sections with holes for fasteners.

Clause 6.2.1(4)

Clause 6.2.1(5)

Clause 6.2.1(4) allows the resistance of all cross-sections to be verified elastically (provided effective properties are used for class 4 sections). For this purpose, the familiar von Mises yield criterion is offered in *clause 6.2.1(5)*, as given by *expression 6.15*, whereby the interaction of the local stresses (divided by the partial factor γ_{M1}) should not exceed the yield stress with more than a constant $C \geq 1$ at any critical point:

$$\left(\frac{\sigma_{x,Ed}}{f_o/\gamma_{M1}}\right)^2 + \left(\frac{\sigma_{z,Ed}}{f_o/\gamma_{M1}}\right)^2 - \left(\frac{\sigma_{x,Ed}}{f_o/\gamma_{M1}}\right)\left(\frac{\sigma_{z,Ed}}{f_o/\gamma_{M1}}\right) + 3\left(\frac{\tau_{Ed}}{f_o/\gamma_{M1}}\right)^2 \leq C \qquad (6.15)$$

$$\frac{\sigma_{x,Ed}}{f_o/\gamma_{M1}} \leq 1 \qquad (6.15a)$$

$$\frac{\sigma_{z,Ed}}{f_o/\gamma_{M1}} \leq 1 \qquad (6.15b)$$

$$\frac{\sqrt{3}\,\tau_{Ed}}{f_o/\gamma_{M1}} \leq 1 \qquad (6.15c)$$

where

$\sigma_{x,Ed}$ is the design value of the local longitudinal stress at the point of consideration
$\sigma_{z,Ed}$ is the design value of the local transverse stress at the point of consideration
τ_{Ed} is the design value of the local shear stress at the point of consideration
C is a constant ≥ 1. The recommended value in Eurocode 9 is $C = 1.2$.

The constant $C > 1$ means that some partially plastic strain is allowed locally for the combination of stresses. However, for the individual stresses no plastic strains are allowed according to *expressions 6.15a, 6.15b* and *6.15c*). Note that the constant C in *criterion 6.15* may be defined in the National Annex, and it may have a different value in some European countries.

Although *equation 6.15* is provided, the majority of design cases can be more efficiently and effectively dealt with using the interaction expressions given throughout *Section 6* of the code, since these are based on the readily available member forces and moments, and they allow more favourable (plastic or partially plastic) interactions.

6.2.2 Section properties

General

Clause 6.2.2

Clause 6.2.2 covers the calculation of cross-sectional properties. Provisions are made for the determination of gross and net areas, effective properties for sections susceptible to shear lag, HAZ softening and local buckling (class 4 cross-section parts).

Gross and net areas

Eurocode 9 can be somewhat confusing, as the term gross area (A_g) used in most clauses is based on nominal dimensions less deductions for HAZ softening due to welds rather than the usual convention of being based on nominal dimensions only. No reduction to the gross area is made for fastener holes, but allowance should be made for larger openings, such as those for services. Note that Eurocode 9 uses the generic term 'fasteners' to cover bolts, rivets and pins.

The net area of the cross-section is taken as the gross area less appropriate deductions for fastener holes, other openings and HAZ softening.

For a non-staggered arrangement of fasteners, for example as shown in Figure 6.4(a), the total area to be deducted should be taken as the sum of the sectional areas of the holes on any line (1–1) perpendicular to the member axis that passes through the centreline of the holes.

For a staggered arrangement of fasteners, for example as shown in Figure 6.4(b), the total area to be deducted should be taken as the greater of:

Figure 6.4. (a) Non-staggered arrangement of fasteners, (b) staggered arrangement of fasteners, (c) angle with holes in both legs

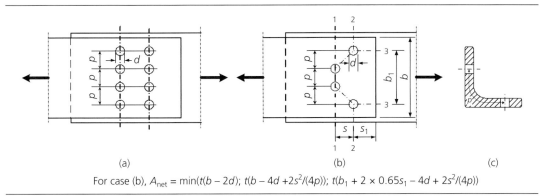

For case (b), $A_{net} = \min(t(b - 2d);\ t(b - 4d + 2s^2/(4p));\ t(b_1 + 2 \times 0.65s_1 - 4d + 2s^2/(4p))$

- the maximum sum of the sectional areas of the holes on any line (1–1) perpendicular to the member axis
- a deduction taken as $\sum td - \sum tb_s$, where b_s is the lesser of

$$s^2/4p \qquad \text{or} \qquad 0.65s \tag{6.16}$$

measured on any diagonal or zig-zag line (2–2) or (3–3), where

- d is the diameter of a hole
- s is the staggered pitch, the spacing of the centres of two consecutive holes in the chain measured parallel to the member axis
- p is the spacing of the centres of the same two holes measured perpendicular to the member axis
- t is the thickness (or effective thickness in a member containing HAZ material).

Clause 6.2.2.2(5) states that for angles or other members with holes in more than one plane, the spacing p should be measured along the centre of thickness of the material (as shown in Figure 6.4(c)). With reference to the figure, the spacing p therefore comprises two straight portions and one curved portion of radius equal to the root radius plus half of the material thickness.

Clause 6.2.2.2(5)

Effective areas to account for local buckling, HAZ and shear lag effects

Eurocode 9 employs an effective area concept to take account of local plate buckling (for slender compression elements), HAZ effects (for longitudinally welded sections) and the effects of shear lag (for wide flanges with low in-plane stiffness).

To distinguish between losses of effectiveness due to local buckling and HAZ on one side and due to shear lag on the other side (and due to a combination of the three effects), Eurocode 9 applies the following effective thicknesses and effective width:

- the effective thickness $t_{eff} = \rho_c t$ is used in relation to local plate buckling effects
- the effective thickness $t_{eff} = \rho_{o,haz}t$ is used in relation to HAZ effects of longitudinal welds
- the effective thickness $t_{eff} = \min(\rho_c t, \rho_{o,haz}t)$ is used due to a combination of the local buckling and HAZ effects (of longitudinal welds) within b_{haz}
- the effective width $b_{eff} = \beta_s b_0$ is used in relation to shear lag effects.

The effective thickness concept is given in ***clause 6.1.5*** (local buckling) and ***clause 6.1.6*** (HAZ), and the effective width (shear lag) in *Annex K* (see Section 9.10 of this guide).

Clause 6.1.5
Clause 6.1.6

The effect of local transverse welds is allowed for by the reduction factor $\omega_{x,haz}$ or $\omega_{xLT,haz}$ (see ***clause 6.3.3.3***), not by using the effective area.

Clause 6.3.3.3

Simple method to take local buckling into account

Due to the low elastic modulus, deflections at the serviceability limit state are often decisive in the design of aluminium structures. When design is based on deflections, it may not be necessary to calculate the resistance exactly, and simple conservative methods are sufficient. The method may be used to generate a quick, approximate and safe solution, perhaps for the purpose of initial member sizing, with the opportunity to refine the calculation for final design.

The following method is not given in Eurocode 9, but as it is conservative it may be used to simplify calculations considerably because no effective cross-section needs to be built up.

The cross-section resistance under axial compression and bending moment may be given by

$$N_{Rd} = \rho_{min} A_g f_o / \gamma_{M1} \tag{D6.3}$$

$$M_{Rd} = \rho_{min} W_{el} f_o / \gamma_{M1} \tag{D6.4}$$

where:

ρ_{min} is the reduction factor for the most slender part of the cross-section (e.g. the part with the largest value of β/β_3)
A_g is the area of the gross cross-section
W_{el} is the section modulus of the gross cross-section.

Only the slenderness β for the cross-section parts in compression and only one limit β_3 (or two, if there are both interior and outstand cross-section parts) need to be calculated. Furthermore, the cross-section class does not need to be calculated. If $\beta < \beta_3$ for all cross-section parts, then $\rho_{min} = 1$.

If there are sections with holes, then net section resistance needs to be checked.

If the section is welded, then $\rho_{min} = \rho_{o,haz}$ may be used. However, using this value may be very conservative, but still sufficient in many cases.

The simple method may also be formulated in a way that is more familiar to many designers. The stress σ_{Ed} can be calculated according to elastic theory and compared with a 'permissible' stress

$$\sigma_{Rd} = \rho_{min} f_o / \gamma_{M1} \tag{D6.5}$$

using the verification $\sigma_{Ed} \leq \sigma_{Rd}$.

If there is no local buckling risk, then σ_{Rd} according to Equation D6.5 is clearly f_o / γ_{M1}.

Shear lag

Calculation of effective widths for wide flanges susceptible to shear lag is covered in *Annex K*, where it states that shear lag effects in flanges may be neglected provided that the flange width $b_0 < L_e/50$, where L_e is the length between points of zero bending moment. The flange width b_0 is defined as either the width of the outstand (measured from the centreline of the web to the flange tip) or half of the width of an internal element (taken as half of the width between the centrelines of the webs). At the ultimate limit state, the limits are relaxed since there will be some plastic redistribution of stresses across the flange, and shear lag may be neglected if $b_0 < L_e/25$ for support regions and $b_0 < L_e/15$ for sagging bending regions. Since shear lag effects rarely arise in conventional building structures, no further discussion on the subject will be given herein.

The effects of shear lag on plate buckling may be taken into account in the following way:

1   reduce the flange width to an effective width b_{eff} for shear lag as defined for the serviceability limit state
2   reduce the thickness to an effective thickness for local buckling based on the slenderness $\beta = b_{eff}/t$ of the effective width according to 1.

Table 6.7. Summary of formulae for cross-section resistances

	Tension	Compression	Bending
No welds	$N_{o,Rd} = A_g f_o / \gamma_{M1}$ (6.18)	$N_{c,Rd} = A_{eff} f_o / \gamma_{M1}$ (6.22)	$M_{c,Rd} = \alpha W_{el} f_o / \gamma_{M1}$ (6.25)
Longitudinal welds	$N_{o,Rd} = A_g f_o / \gamma_{M1}$ (6.18)	$N_{c,Rd} = A_{eff} f_o / \gamma_{M1}$ (6.22)	$M_{c,Rd} = \alpha W_{el} f_o / \gamma_{M1}$ (6.25)
Filled bolt holes		$N_{c,Rd} = A_{eff} f_o / \gamma_{M1}$ (6.22)	$M_{u,Rd} = W_{net} f_u / \gamma_{M2}$ (6.24)
Unfilled, oversized or slotted holes	$N_{u,Rd} = 0.9 A_{net} f_u / \gamma_{M2}$ (6.19a)	$N_{u,Rd} = A_{net} f_u / \gamma_{M2}$ (6.21)	
Transverse welds	$N_{u,Rd} = A_{u,eff} f_u / \gamma_{M2}$ (6.19b)	$N_{u,Rd} = A_{u,eff} f_u / \gamma_{M2}$ (6.21b)	$M_{u,Rd} = W_{u,eff,haz} f_u / \gamma_{M2}$ (6.24b)

Overview of formulae for cross-section resistance

Formulae for cross-section resistances for tension, compression and bending are summarised in Table 6.7. In general:

- the 0.2% proof strength (f_o) is used for overall yielding
- the 0.2% proof strength in HAZ ($f_{o,haz} = \rho_{o,haz} f_o$) is used for longitudinal welds in profiles
- the ultimate strength in HAZ ($f_{u,haz} = \rho_{u,haz} f_u$) is used for transverse (localised) welds
- the ultimate strength (f_u) is used in sections with holes.

A_g is either the gross section or a reduced cross-section to allow for HAZ softening due to longitudinal welds. In the latter case, A_g is found by taking a reduced area equal to $\rho_{o,haz}$ times the area of the HAZ (see *clause 6.1.6.2*).

Clause 6.1.6.2

A_{net} is the net section area, with a deduction for holes and a deduction, if required, to allow for the effect of HAZ softening in the net section through the hole. The latter deduction is based on the reduced thickness of $\rho_{u,haz} t$.

A_{eff} is the effective cross-section area, obtained using a reduced thickness $\rho_c t$ for class 4 parts and a reduced thickness $\rho_{o,haz} t$ for the HAZ material, whichever is smaller, but ignoring unfilled holes.

$A_{u,eff}$ is the effective cross-section in a section with transverse welds. For tension members, $A_{u,eff}$ is based on the reduced thickness $\rho_{u,haz} t$. For compression members, $A_{u,eff}$ is the effective section area, obtained using a reduced thickness $\rho_c t$ for class 4 parts and a reduced thickness $\rho_{u,haz} t$ for the HAZ material, whichever is smaller.

α is the shape factor (see *clause 6.2.5*).

Clause 6.2.5

W_{el} is the elastic modulus of the gross cross-section.

W_{net} is the elastic modulus of the net section, allowing for holes and HAZ softening, if welded (see Section 6.2.5). The latter deduction is based on the reduced thickness of $\rho_{u,haz} t$.

$W_{u,eff,haz}$ is the effective section modulus, obtained using a reduced thickness $\rho_c t$ for class 4 parts and a reduced thickness $\rho_{u,haz} t$ for the HAZ material, whichever is smaller.

The cross-section resistances are further explained later in this guide.

6.2.3   Tension

The resistance of tension members is covered in *clause 6.2.3*. The design tensile force is denoted by N_{Ed} (axial design effect). The tensile design resistance is limited either by yielding $N_{o,Rd}$ of the gross cross-section (to prevent excessive deformation of the member) or ultimate failure $N_{u,Rd}$ of the net cross-section (at holes for fasteners) or ultimate failure at the section with localised HAZ softening, whichever is the lesser:

Clause 6.2.3

■ general yielding along the member:

$$N_{o,Rd} = A_g f_o / \gamma_{M1} \qquad (6.18)$$

■ local failure at a section with holes:

$$N_{u,Rd} = 0.9 A_{net} f_u / \gamma_{M2} \qquad (6.19a)$$

■ local failure at a section with transverse weld:

$$N_{u,Rd} = A_{u,eff} f_u / \gamma_{M2} \qquad (6.19b)$$

where the notation follows Table 6.7.

Note that any increased eccentricity due to a shift of centroid axis arising from reduced sections of HAZ may be neglected.

Clause 8.5.2.3

For angles connected through one leg, see *clause 8.5.2.3*. Similar consideration should also be given to other types of sections connected through outstands such as T sections and channels.

Clause 6.2.2

For staggered holes, see *clause 6.2.2*.

Example 6.1: tension resistance of a bar with bolt holes and an attachment

Two extruded flat bars ($b = 150$ mm wide and $t = 5$ mm thick) are to be connected together with a lap splice using six M12 bolts, as shown in Figure 6.5. Calculate the tensile strength, assuming EN AW-6063 T6, which, according to Table 3.1, has a proof strength $f_o = 160$ MPa. The partial factors of strength are $\gamma_{M1} = 1.1$ and $\gamma_{M2} = 1.25$ according to Table 6.1.

A 100 mm plate is attached to one of the bars by MIG welding. How much is the resistance reduced by the attachment?

Gross area resistance (without welds)

$$N_{o,Rd} = A_g f_o / \gamma_{M1} = 150 \times 5 \times 250 / 1.1 = 170 \, kN$$

Net section resistance

The net sections in lines 1 and 2 and minimum resistance are

$$A_{net,1} = A_g - td = 150 \times 5 - 5 \times 12 = 690 \, mm^2$$

Figure 6.5. Lap splice in a tension member with staggered bolts and a welded attachment

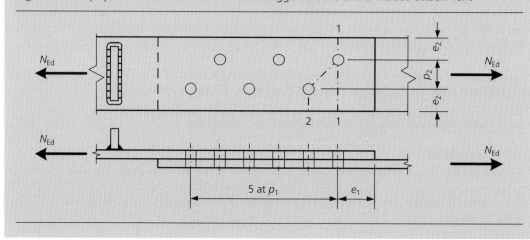

$$A_{\text{net,2}} = A_g - (2td - 2ts^2/4p) = 150 \times 5 - [2 \times 5 \times 12 - 2 \times 5 \times 75^2/(4 \times 75)] = 818\,\text{mm}^2$$

$$N_{\text{u,Rd}} = 0.9 A_{\text{net,1}} f_u / \gamma_{\text{M2}} = 0.9 \times 690 \times 290/1.25 = 144\,\text{kN}$$

The tensile resistance in the splice is

$$N_{\text{t,Rd}} = \min(N_{\text{o,Rd}}, N_{\text{u,Rd}}) = 144\,\text{kN}$$

Resistance in the HAZ
As the width of the HAZ is 20 mm (see Table 6.6), it covers almost the whole width of the cross-section. The reduction factor in the HAZ is, according to *Table 3.2*, $\rho_{\text{u,haz}} = 0.64$.

$$A_{\text{u,eff}} = t(b - b_a - 2b_{\text{haz}}) + \rho_{\text{u,haz}} t(b_a + 2b_{\text{haz}}) = 5(150 - 140) + 0.64 \times 5 \times 140 = 498\,\text{mm}^2$$

$$N_{\text{u,Rd}} = A_{\text{u,eff}} f_u / \gamma_{\text{M2}} = 498 \times 290/1.25 = 116\,\text{kN}$$

The welded attachment reduces the resistance by 20%.

6.2.4   Compression

Cross-section resistance in compression is covered in *clause 6.2.4*. This ignores overall member buckling effects, and therefore may only be applied as the sole check to a member of low slenderness ($\bar{\lambda} \leq 0.2$). For all other cases, checks also need to be made for member buckling as defined in *clause 6.3*.

Clause 6.2.4

Clause 6.3

A simple conservative method to allow for local buckling and HAZ effects is given in Section 6.2.2 of this guide.

The design compressive force is denoted by N_{Ed} (the axial design effect). The design resistance of a cross-section under uniform compression, $N_{\text{o,Rd}}$ is the lesser of

- in sections with unfilled holes

$$N_{\text{u,Rd}} = A_{\text{net}} f_u / \gamma_{\text{M2}} \qquad (6.21)$$

- in sections with transverse weld

$$N_{\text{u,Rd}} = A_{\text{u,eff}} f_u / \gamma_{\text{M2}} \qquad (6.21b)$$

- other sections

$$N_{\text{o,Rd}} = A_{\text{eff}} f_o / \gamma_{\text{M1}} \qquad (6.22)$$

where the notation follows Table 6.7.

Note that any increased eccentricity due to a shift of the centroid axis arising from reduced sections of the HAZ may be neglected.

Example 6.2: resistance of an I cross-section in compression
An extruded profile is to be used as a short compression member (Figure 6.6). Calculate the resistance of the cross-section in compression using the material EN AW-6082 T6.

Section properties

Section height	$h = 200\,\text{mm}$
Flange width	$b = 100\,\text{mm}$
Flange thickness	$t_f = 9\,\text{mm}$
Web thickness	$t_w = 6\,\text{mm}$

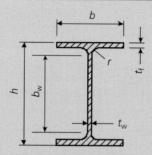

Figure 6.6. Cross-section notation

Fillet radius	$r = 14$ mm
Web height	$b_w = h - 2t_f - 2r = 154$ mm
EN AW-6082 T6	$f_o = 260$ MPa
Partial safety factor	$\gamma_{M1} = 1.1$

Clause 6.1.4

Cross-section classification under axial compression (*clause 6.1.4*)

$$\varepsilon = \sqrt{250/f_o} = \sqrt{250/260} = 0.981$$

Outstand flanges (*expression 6.1*):

$$\beta_f = (b - t_w - 2r)/2t_f = (100 - 6 - 2 \times 14)/(2 \times 9) = 3.67$$

Limits for classes 1 and 2:

$$\beta_1 = 3\varepsilon = 2.94 < \beta_f$$

$$\beta_2 = 4.5\varepsilon = 4.41 > \beta_f$$

The flange is class 2.

Web – internal part in compression (*expression 6.1*):

$$\beta_w = b_w/t_w = 154/6 = 25.7$$

Limits for class 3:

$$\beta_3 = 22\varepsilon = 21.6 < \beta_w$$

The web is class 4.

In compression, the overall cross-section classification is class 4. The resistance is therefore based on the effective cross-section for the member in compression.

Clause 6.3.1

Cross-section compression resistance (*clause 6.3.1*)
To calculate the effective cross-section area, the gross cross-section area is first calculated and then the reduction due to local buckling is made. The fillets are included.

$$A_g = bh - (b - t_w)(h - 2t_f) + r^2(4 - \pi) = 100 \times 200 - 94 \times 182 + 14^2(4 - \pi)$$

$$= 3060 \text{ mm}^2$$

Web slenderness according to the above:

$$\beta_w = b_w/t_w = 154/6 = 25.7$$

$$\beta_w/\varepsilon = 25.7/0.981 = 26.2$$

Reduction factor (*clause 6.1.5*) with $C_1 = 32$ and $C_2 = 220$ from Table 6.5 (*Table 6.3*) class A, no weld:

Clause 6.1.5

$$\rho = \frac{C_1}{\beta/\varepsilon} - \frac{C_2}{(\beta/\varepsilon)^2} = \frac{32}{26.2} - \frac{220}{26.2^2} = 0.901 \tag{6.12}$$

$$A_{\mathrm{eff}} = A_{\mathrm{g}} - b_{\mathrm{w}}(t_{\mathrm{w}} - \rho t_{\mathrm{w}}) = 3060 - 154(6 - 0.901 \times 6) = 2969\,\mathrm{mm}^2$$

Cross-section compression resistance:

$$N_{\mathrm{Rd}} = A_{\mathrm{eff}} f_{\mathrm{o}}/\gamma_{\mathrm{M1}} = 2969 \times 260/1.1 = 702\,\mathrm{kN} \tag{6.22}$$

Example 6.3: resistance of a class 4 hollow section in compression

Aluminium profiles may have very different and complicated shapes. Examples of profiles used in curtain walls and windows are shown in Figure 6.7. The cross-section may have bolt channels and screw ports that may work as stiffeners of slender parts of the cross-section. The fourth profile is chosen as an example of a class 4 cross-section for axial load.

The aim is to calculate the resistance of the profile in Figure 6.8 in compression. The material is EN AW-6063 T6, which, according to Table 3.1, belong to buckling class A, and has a proof strength $f_{\mathrm{o}} = 160\,\mathrm{MPa}$. The partial factor of strength is $\gamma_{\mathrm{M1}} = 1.1$ according to Table 6.1, and the modulus of elasticity is 70 000 MPa according to *clause 3.2*.

Clause 3.2

The cross-section is complicated. Usually, the 'ordinary' cross-section constants such as the cross-section area, second moment of the area and the section modulus can be obtained from a CAD program, which is used in this example.

The cross-section constants needed are found to be $A = 856.1\,\mathrm{mm}^2$, $I_{\mathrm{y}} = 1.184 \times 10^6\,\mathrm{mm}^4$, $W_{\mathrm{y,el}} = 1.959 \times 10^4\,\mathrm{mm}^3$ and $z_{\mathrm{gc}} = 56.55\,\mathrm{mm}$. Some measurements are also needed to check local buckling: $b = 50\,\mathrm{mm}$, $t_{\mathrm{f}} = 2\,\mathrm{mm}$, $h = 100\,\mathrm{mm}$ and $t_{\mathrm{w}} = 2\,\mathrm{mm}$ (see Figure 6.8). Furthermore, for local and distortional buckling resistance of the webs, $s_1 = 84.9\,\mathrm{mm}$ and $s_2 = 38.0\,\mathrm{mm}$ (measured from the midpoint between the top flange and the small 'stiffener' close to the bottom flange) and $b_1 = 26.4\,\mathrm{mm}$ and $b_2 = 36.2\,\mathrm{mm}$ are needed.

The web stiffeners (screw ports) close to the centre of the webs have a noticeable influence on the axial force resistance. Three methods in Eurocode 9 may be used.

Figure 6.7. Examples of typical aluminium profiles for curtain walls and windows

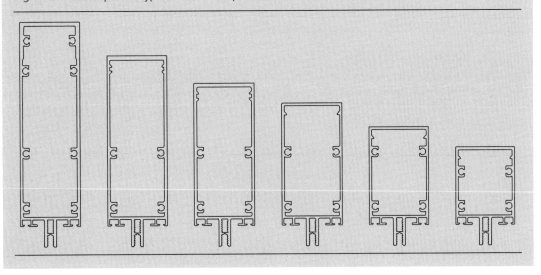

Figure 6.8. (a) Aluminium profile with screw ports and bolt channels, (b) cross-section for the second moment of the area of the web stiffeners, (c) cross-section for the effective area of the web stiffeners and (d) the effective cross-section

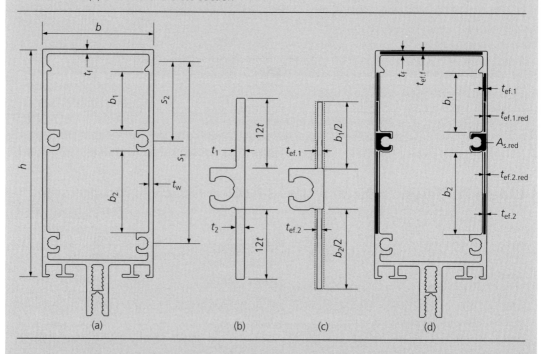

Clause 6.1.4.3

1 the diagram in *clause 6.1.4.3*
2 the procedure in EN 1999-1-4 for trapezoidal sheeting

Clause 6.7.3

3 the procedure in *clause 6.7.3*.

The first method is only applicable if the stiffener is located in the middle of the web. The second method is more general, and is therefore used in this example. The third is essentially the same as the second, except that it is meant to be used for plate girders with transverse stiffeners as well. The reduction factor for distortional buckling of the stiffener is therefore the same as for flexural (column) buckling between transverse stiffeners, so method 3 is too conservative in this case.

The screw ports at the bottom flange and the small stiffeners at the top flange are taken account of by reducing the web depth. The bottom flange is so stiffened that it needs not to be checked for local buckling.

Clause 6.1.4

Cross-section classification (*clause 6.1.4*)
In Table 6.4 (*Table 6.2*),

$$\varepsilon = \sqrt{250/160} = 1.25$$

Clause 6.1.4.3(1)

Flange (*clause 6.1.4.3(1)*):

$$\beta_f = (b - 2t_w)/t_f = (50 - 2 \times 2)/2 = 23$$

Limits in Table 6.4 (*Table 6.2*):

$$\beta_1 = 16\varepsilon = 11 \times 1.25 = 13.8$$

$$\beta_2 = 16\varepsilon = 16 \times 1.25 = 20$$

$$\beta_3 = 22\varepsilon = 22 \times 1.25 = 27.5$$

The flange is class 3.

Upper and lower part of the web (*clause 6.1.4.3(1)*):

Clause 6.1.4.3(1)

$$\beta_w = b_1/t_w = 26.4/2 = 13.2$$

The web part 1 is class 1.

$$\beta_w = b_2/t_w = 36.2/2 = 18.1$$

The web part 2 is class 2.

As the cross-section class is ≤ 3, there is no reduction due to local buckling of the flat parts between the stiffeners ($t_{ef,1} = t_{ef,2} = t_w$). However, distortional buckling of the web stiffener near the centre of the web may reduce the resistance. For distortional buckling, the stiffener is working as a compressed strut on an elastic foundation according to EN 1999-1-4, *clause 5.5.4.3* (see Section 10.5.3 of this guide). The second moment of the area and the area of the stiffener and the adjacent flat part of the web according to Figures 6.8(b) and 6.8(c) are required. These are found using the CAD program:

*EN 1999-1-4:
Clause 5.5.4.3*

$$I_{sa,ef} = 428 \text{ mm}^4$$

$$A_{sa,ef} = 103 \text{ mm}^2$$

*EN 1999-1-4:
Clause 5.5.4.3*

The buckling stress is given by *expression 5.23* in EN 1999-1-4, *clause 5.5.4.3* with the factor $\kappa_f = 1.0$. (See also Sections 10.5.3 and 10.5.4 of this guide.)

$$\sigma_{cr,sa} = \frac{1.05\kappa_f E}{A_{sa,ef}} \sqrt{\frac{I_{sa,ef}\, t_w^3 s_1}{s_2(s_1 - s_2)}}$$

$$= \frac{1.05 \times 1.0 \times 70\,000}{103} \sqrt{\frac{428 \times 2^3 \times 84.9}{38.0(84.9 - 38.0)}} = 216 \text{ MPa} \qquad (5.23)$$

*EN 1999-1-4:
Clause 5.5.3.1*

The slenderness is given by *expression 5.7* in EN 1999-1-4, *clause 5.5.3.1*. (See also Section 10.5.5 of this guide.)

$$\bar{\lambda}_s = \sqrt{\frac{f_o}{\sigma_{cr,sa}}} = \sqrt{\frac{160}{216}} = 0.861 \qquad (5.7)$$

and the reduction factor is given by the expression in *Table 5.4* in the same clause:

$$\chi_r = 1.155 - 0.62\bar{\lambda}_s = 1.155 - 0.62 \times 0.861 = 0.621 \qquad \text{for } 0.25 < \bar{\lambda}_s < 1.04$$

The effective area can now be found by drawing the effective cross-section in the CAD program using the effective thickness:

$$t_{ef,1,red} = \chi_r t_{ef,1} = 0.621 \times 2.0 = 1.24 \text{ mm}$$

$$t_{ef,2,red} = \chi_r t_{ef,2} = 0.621 \times 2.0 = 1.24 \text{ mm}$$

and also reducing the area of the stiffener (screw port) itself with the factor 0.621.

Cross-section compression resistance

Clause 6.2.4

The resulting effective area is $A_{eff} = 778 \text{ mm}^2$, and the axial force resistance according to *expression 6.22* is

$$N_{o,Rd} = A_{eff} f_o / \gamma_{M1} = 778 \times 160/1.1 = 113 \text{ kN} \qquad (6.22)$$

Figure 6.9. Situations where lateral torsional buckling may be ignored (except (b1))

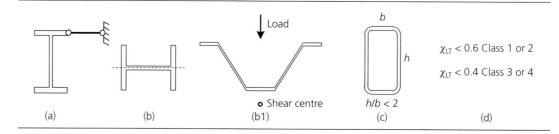

6.2.5. Bending moment

Clause 6.2.5

Clause 6.3.2

Cross-section resistance in bending is covered in **clause 6.2.5**, and represents the in-plane flexural strength of a beam with no account taken of lateral torsional buckling. The lateral torsional buckling check is described in **clause 6.3.2**. There are many situations where lateral torsional buckling may be ignored (Figure 6.9):

(a) where sufficient lateral restraint is applied to the compression flange of the beam
(b) where bending is about the minor axis of symmetric sections (unless the load application point is above the shear centre, Figure 6.9(b1))
(c) where cross-sections with high lateral and torsional stiffness are employed, for example rectangular (height over width less than 2) and circular hollow sections
(d) generally where the slenderness for lateral torsional buckling is less than the limit $\bar{\lambda}_{0,\mathrm{LT}}$ of the horizontal plateau for the reduction factor χ_{LT} for lateral torsional buckling ($\bar{\lambda}_{0,\mathrm{LT}} = 0.4$ for class 3 and 4 cross-sections and $\bar{\lambda}_{0,\mathrm{LT}} = 0.6$ for class 1 and 2 cross-sections: see **clause 6.3.2.2**).

Clause 6.3.2.2

The design bending moment is denoted by M_{Ed} (the bending moment design effect). The design resistance M_{Rd} for bending about one principal axis of a cross-section is determined as the lesser of $M_{\mathrm{u,Rd}}$ and $M_{\mathrm{o,Rd}}$, where:

$$M_{\mathrm{u,Rd}} = W_{\mathrm{net}} f_{\mathrm{u}}/\gamma_{\mathrm{M2}} \qquad \text{in a net section} \tag{6.24}$$

$$M_{\mathrm{u,Rd}} = W_{\mathrm{u,eff,haz}} f_{\mathrm{u}}/\gamma_{\mathrm{M2}} \qquad \text{at a section with localised transverse weld} \tag{6.24b}$$

$$M_{\mathrm{o,Rd}} = \alpha W_{\mathrm{el}} f_{\mathrm{o}}/\gamma_{\mathrm{M1}} \qquad \text{at each cross-section} \tag{6.25}$$

where:

Clause 6.2.5.1

α is the shape factor (see Table 6.8 (*Table 6.4* in **clause 6.2.5.1**))
W_{el} is the elastic modulus of the gross section
W_{net} is the elastic modulus of the net section allowing for holes and HAZ softening, if welded (the latter deduction is based on the reduced thickness of $\rho_{\mathrm{o,haz}}t$)
$W_{\mathrm{u,eff,haz}}$ is the effective section modulus, obtained using a reduced thickness $\rho_{\mathrm{c}}t$ for class 4 parts and a reduced thickness $\rho_{\mathrm{u,haz}}t$ for the HAZ material, whichever is smaller.

In *expressions 6.26* and *6.27*, β is the slenderness for the most critical part of the section, and β_2 and β_3 are the limiting values for that same part according to Table 6.4 (*Table 6.2* of EN 1999-1-1). The critical part is determined by the lowest value of $(\beta_3 - \beta)/(\beta_3 - \beta_2)$. Additionally:

W_{pl} is the plastic modulus of the gross section
$W_{\mathrm{pl,haz}}$ is the plastic modulus of the gross section, obtained using a reduced thickness $\rho_{\mathrm{o,haz}}t$ for the HAZ material
$W_{\mathrm{el,haz}}$ is the effective elastic modulus of the gross section, obtained using a reduced thickness $\rho_{\mathrm{o,haz}}t$ for the HAZ material
W_{eff} is the effective section modulus, obtained using a reduced thickness t_{eff} for the class 4 parts
$W_{\mathrm{eff,haz}}$ is the effective section modulus, obtained using a reduced thickness t_{eff} for the class 4 parts and a reduced thickness $\rho_{\mathrm{c}}t$ or $\rho_{\mathrm{o,haz}}t$ for the HAZ material, whichever is the smaller.

Table 6.8. Shape factor α (*clause 6.2.5.1*)

Clause 6.2.5.1

Cross-section class	Without welds	With longitudinal welds
1	$\alpha_1 \geq W_{pl}/W_{el}$ See *Annex F* (Section 9.6 of this guide)	$\alpha_1 \geq W_{pl,haz}/W_{el}$ See *Annex F* (Section 9.6 of this guide)
2	$\alpha_2 = W_{pl}/W_{el}$	$\alpha_2 = W_{pl,haz}/W_{el}$
3	$$\alpha_{3,u} = \left[1 + \left(\frac{\beta_3 - \beta}{\beta_3 - \beta_2}\right)\left(\frac{W_{pl}}{W_{el}} - 1\right)\right]$$ (6.26) or $\alpha_{3,u} = 1$ for simplicity	$$\alpha_{3,w} = \left[\frac{W_{el,haz}}{W_{el}} + \left(\frac{\beta_3 - \beta}{\beta_3 - \beta_2}\right)\left(\frac{W_{pl,haz} - W_{el,haz}}{W_{el}}\right)\right]$$ (6.27) or $\alpha_{3,w} = \dfrac{W_{el,haz}}{W_{el}}$ for simplicity
4	$\alpha_4 = W_{eff}/W_{el}$	$\alpha_4 = W_{eff,haz}/W_{el}$

Reproduced from EN 1999-1-1 (*Table 6.4*), with permission from BSI.

The effective cross-section of two class 4 compression flanges (and parts of webs) of welded members are illustrated in Figure 6.10 for two cases: $\rho_{o,haz} > \rho_c$ and $\rho_{o,haz} < \rho_c$.

For a welded part in class 3 or 4 sections, a more favourable assumed thickness may be taken as given in *clause 6.2.5.2(2)e*.

Clause 6.2.5.2(2)e

Generally, the calculation of the effective thickness of the web requires an iterative procedure, as the reduction in the web thickness is dependent on the position of the neutral axis, which is changed when the web is reduced. However, in *clause 6.1.4.4(4)*, a simplified approach is allowed that ends in two steps. In the first step, the effective thickness of the flange (if it is in class 4) is determined from the stress distribution of the gross cross-section. In the second step, the stresses are determined based on the cross-section composed of the effective area of the compression flanges and the gross area of the web and the tension flange. The effective thickness of the web is calculated based on these stresses, and this is taken as the final result. The procedure is illustrated in the following examples.

Clause 6.1.4.4(4)

Effective cross-section for a symmetric I girder with class 1, 2 or 3 flanges and a class 4 web

The effective parts of a class 4 cross-section are combined into an effective cross-section. An example of the effective cross-section for a symmetric I girder with class 1, 2 or 3 flanges and a class 4 web is given in Figure 6.11. Note that no iteration process is necessary. The effective thickness of the web is based on the width b_w and $\psi = -1.0$. The web thickness is reduced to

Figure 6.10. Effective cross-section of class 4 compression parts of welded members

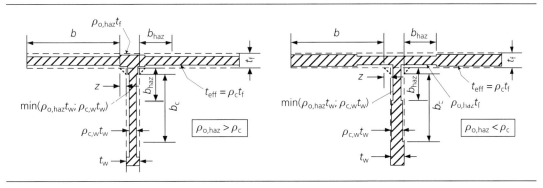

Figure 6.11. Effective cross-section for a symmetrical member: (a) the gross cross-section; (b) the effective cross-section for the bending moment

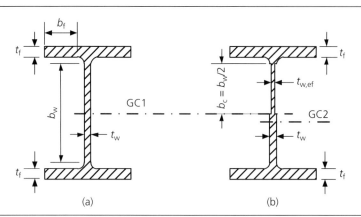

the effective thickness $t_{w,ef}$ within $b_c = b_w/2$ on the compression side only. GC1 is the centre of gravity for the gross cross-section, and GC2 is the centre of gravity for the effective section.

Symmetric I girder with class 4 flanges and a web or an I girder with different flanges
Calculation steps for a symmetric I girder with class 4 flanges and a web, or an asymmetric I girder (different flange area) with a class 4 web are given in Figure 6.12. The effective thickness of the web is based on the width b_w and $\psi = 1 - b_w/b_c$, where b_c is the width of the compression part of the web calculated for the cross-section with a reduced compression flange but an unreduced web (step 1). The web thickness is reduced to the effective thickness $t_{w,ef}$ within b_c (step 2).

Members containing localised welds
If a section is affected by HAZ softening with a specified location along the length and if the softening does not extend longitudinally a distance greater than the least width of the member, then the limiting stress should be taken as the design **ultimate strength** $\rho_{u,haz} f_u / \gamma_{M2}$ of the reduced strength material (***clause 6.2.9.3***). Remember that welding of temporary attachments also results in HAZ effects.

Clause 6.2.9.3

If the softening extends longitudinally a distance greater than the least width of the member, the limiting stress should be taken as the strength $\rho_{o,haz} f_o$ for **overall yielding** of the reduced-strength material.

In a longitudinally welded member with a localised (transverse) weld, the ultimate strength may be used for all welds in the section. See Example 6.7.

Figure 6.12. Effective cross-section for a welded member: (a) step 1, the reduced flange; (b) step 2, the effective cross-section

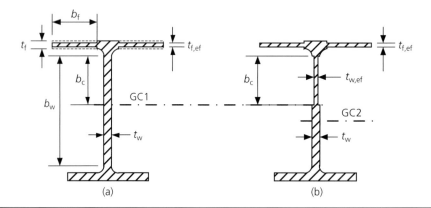

Simple method to take local buckling and HAZ softening into account
As already pointed out in Section 6.2.2, due to the low elastic modulus, deflections at the service-ability limit state are often decisive in the design of aluminium structures. It is then not necessary to calculate the resistance exactly, and the simple conservative method in Section 6.2.2 may be sufficient.

Example 6.4: bending moment resistance of a class 1 cross-section
An extruded member is to be designed in bending (Figure 6.13). The proportions of the section have been selected in such a way that it may be classified as a class 1 cross-section. The material is EN AW-6005A T6.

Section properties
The cross-section dimensions are:

Section height	$h = 200$ mm
Flange width	$b = 160$ mm
Flange thickness	$t_f = 25$ mm
Web thickness	$t_w = 16$ mm
Web height	$b_w = h - 2t_f = 156$ mm
EN AW-6082 T6	$f_o = 200$ MPa (10 mm $< t \leq 25$ mm)
Partial safety factor	$\gamma_{M1} = 1.1$

Cross-section classification (*clause 6.1.4*)

Clause 6.1.4

$$\varepsilon = \sqrt{250/f_o} = \sqrt{250/200} = 1.12$$

Outstand flanges (*6.1*):

$$\beta_f = (b - t_w)/2t_f = (160 - 16)/(2 \times 25) = 3.27 < \beta_1 = 3\varepsilon = 3.354$$

The flange is class 1.

Web — internal part (*expression 6.1*):

$$\beta_w = 0.4b_w/t_w = 0.4 \times 156/16 = 3.90 < \beta_1 = 11\varepsilon = 12.3$$

The web is class 1.

The overall cross-section classification is class 1. The resistance is therefore plastic bending resistance or a refined value according to *Annex F*.

Bending moment resistance of the cross-section (*clause 6.2.5*)

Clause 6.2.5

The elastic section modulus and the plastic section modulus are (see Figure 6.13)

$$W_{el} = \frac{1}{12}\left(bh^3 - (b - t_w)(h - 2t_f)^3\right)\frac{2}{h} = \frac{1}{12}\left(160 \times 200^3 - 144 \times 156^3\right)\frac{2}{200}$$

$$= 0.611 \times 10^6 \text{ mm}^3$$

$$W_{pl} = bt_f(h - t_f) + \tfrac{1}{4}t_w h_w^2 = 160 \times 22(200 - 22) + \tfrac{1}{4}(16 \times 156^2) = 0.724 \times 10^6 \text{ mm}^3$$

Figure 6.13. Extruded class 1 cross-section

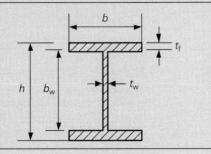

The geometric shape factor is

$$\alpha_0 = W_{pl}/W_{el} = 0.724/0.611 = 1.185$$

If the cross-section was class 2, then $\alpha_2 = \alpha_0$. However, for a class 1 cross-section, the refined generalised shape factor can be used. This generalised shape factor $\alpha_{M,j}$ is found in *Annex F* (*Table F.2*) (see Section 9.5 of this guide), and it depends on the alloy. As the min elongation for EN AW-6005A T6 is 8% according to Table 3.1 (*Table 3.2b*), then α_{10} according to expression D9.2 in Section 9.5 of this guide may be used (see ***clause G(3)***):

Clause G(3)

$$\alpha_{10} = \alpha_0^{[0.21 \log(1000\,n)]} 10^{[7.96 \times 10^{-2} - 8.09 \times 10^{-2} \log(n/10)]} \tag{D9.2}$$

$$\alpha_{10} = 1.185^{0.21 \log(1000 \times 20)} \times 10^{[0.0796 - 0.0809 \log(0.1 \times 20)]} = 1.323$$

The bending moment resistance is

$$M_{Rd} = \alpha_{10} f_o W_{el}/\gamma_{M1} = 1.323 \times 200 \times 0.611 \times 10^6/1.1 = 147\,\text{kN m}$$

The resistance is increased 11.7% by using the generalised shape factor.

Example 6.5: bending moment resistance of a class 3 cross-section

The aim is to calculate the major axis bending moment resistance of the profile in Example 6.3 (see Figure 6.8(a)) for the upper flange in compression. A simplification of the cross-section is shown in Figure 6.14(a). The material is EN AW-6063 T6, which, according to Table 3.1 (*Table 3.2b*), belongs to buckling class A and has a proof strength $f_o = 160$ MPa. The partial factor of strength is $\gamma_{M1} = 1.1$, and the modulus of elasticity is 70 000 MPa.

The cross-section is complicated. From the CAD program, $A = 856.1$ mm^2, $I_y = 1.184 \times 10^6$ mm^4, $W_{y,el} = 1.959 \times 10^4$ mm^3 and $z_{gc} = 56.55$ mm. Also, some measurements are needed to check the local buckling: $b = 46$ mm, $t_f = 2$ mm, $h = 100$ mm, $t_w = 2$ mm, $s_c = 84$ mm and, for the bottom flange, $h_n = 17$ mm (see Figure 6.14(a)).

Figure 6.14. (a) Cross-section (same as Example 6.3 but simplified), (b) stresses and (c) upper half of the cross-section

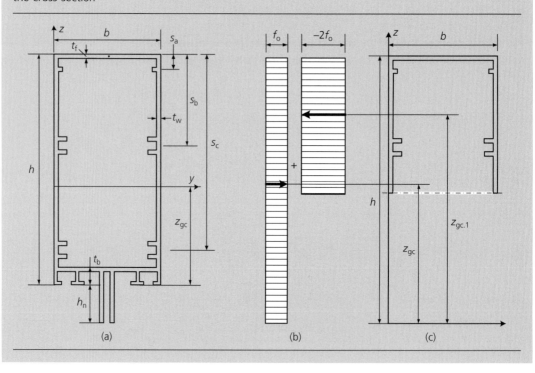

The influence of the web stiffeners (screw ports) close to the centre of the webs is small, and omitted when calculating the major axis moment resistance. For axial force, they may have a noticeable influence: see Example 6.3.

Cross-section classification (*clause 6.1.4*)

In Table 6.4 (*Table 6.2*),

$$\varepsilon = \sqrt{250/160} = 1.25$$

Clause 6.1.4

Flange (*clause 6.1.4.3(1)*):

$$\beta_{\mathrm{f}} = (b - 2t_{\mathrm{w}})/t_{\mathrm{f}} = (46 - 2 \times 2)/2 = 21$$

Clause 6.1.4.3(1)

Limits in Table 6.4 (*Table 6.2*):

$$\beta_2 = 16\varepsilon = 16 \times 1.25 = 20$$

$$\beta_3 = 22\varepsilon = 22 \times 1.25 = 27.5$$

The flange is class 3.

Web (*clause 6.1.4.3(1)*) – distance to the upper edge:

$$z_{\mathrm{ue}} = h + h_{\mathrm{n}} - z_{\mathrm{gc}} = 100 + 17 - 56.55 = 60.45 \,\mathrm{mm}$$

Clause 6.1.4.3(1)

As the tension flange is much stiffened, the web is assumed to start at the middle of the bottom screw port. Then,

$$\psi = -\frac{s_{\mathrm{c}} - z_{\mathrm{ue}}}{z_{\mathrm{ue}} - t_{\mathrm{f}}} = -\frac{84 - 60.45}{60.45 - 2} = -0.403$$

$$\beta_{\mathrm{w}} = (0.7 + 0.3\psi)(s_{\mathrm{c}} - t_{\mathrm{f}})/t_{\mathrm{w}} = [0.7 + 0.3(-0.403)](46 - 2)/2 = 21$$

$$\beta_2 = 16\varepsilon = 16 \times 1.25 = 20$$

$$\beta_3 = 22\varepsilon = 22 \times 1.25 = 27.5$$

The web is class 3

Shape factor

The section classification is class 3, and the shape factor is 1.0, or may alternatively be calculated according to *expression 6.26* (see *clause 6.2.5.1*). As the web area is a large part of the cross-section, *expression 6.26* is used:

Clause 6.2.5.1

$$\alpha_{3,u} = 1 + \left(\frac{\beta_3 - \beta}{\beta_3 - \beta_2}\right)\left(\frac{W_{\mathrm{pl}}}{W_{\mathrm{el}}} - 1\right) \tag{6.26}$$

where the plastic section modulus W_{pl} is needed. Again, the CAD program is used. The cross-section is divided into two parts with the same cross-section area. The plastic section modulus can be calculated as half of the cross-section area times the distance between the centres of gravity of the two halves or the difference between the static moment of two times the upper part around the y–y axis minus the static moment of the whole area. The second method (illustrated in Figure 6.14(b)) gives, with $z_{\mathrm{gc,1}} = 89.19$ mm:

$$W_{\mathrm{pl}} = 2\frac{A}{2}z_{\mathrm{gc,1}} - Az_{\mathrm{gc}} = 856.1 \times (89.19 - 56.55) = 2.794 \times 10^4 \,\mathrm{mm}^3$$

The cross-section part with the smallest value of the ratio $(\beta_3 - \beta)/(\beta_3 - \beta_2)$ is the critical part. For the flange,

$$(\beta_{3\mathrm{f}} - \beta_{\mathrm{f}})/(\beta_{3\mathrm{f}} - \beta_{2\mathrm{f}}) = (27.5 - 21)/(27.5 - 20) = 0.867$$

and for the web,

$$(\beta_{3w} - \beta_w)/(\beta_{3w} - \beta_{2w}) = (27.5 - 23.74)/(27.5 - 20) = 0.501$$

In this case the web is governing, as the limits are the same for the web and the flange, but, in general, $(\beta_3 - \beta)/(\beta_3 - \beta_2)$ needs to be calculated for different parts of the cross-section:

$$\alpha_{3,u} = 1 + \left(\frac{\beta_3 - \beta}{\beta_3 - \beta_2}\right)\left(\frac{W_{pl}}{W_{el}} - 1\right) = 1 + 0.501\left(\frac{2.794 \times 10^4}{1.959 \times 10^4} - 1\right) = 1.214 \qquad (6.26)$$

Bending moment resistance
The bending moment resistance is, from *expression 6.25*,

$$M_{Rd} = \alpha_{3,u}W_{el}f_o/\gamma_{M1} = 1.214 \times 1.959 \times 10^4 \times 160/1.1 = 3.46\,\text{kN m} \qquad (6.25)$$

Example 6.6: bending moment resistance of a class 4 cross-section

The major axis bending moment resistance for the upper flange in compression is to be calculated. The material is EN AW-6063 T6, which, according to Table 3.1, belongs to buckling class A and has a proof strength $f_o = 160$ MPa. The partial factor of strength is $\gamma_{M1} = 1.1$.

The cross-section is complicated. As in the previous example, the 'ordinary' cross-section constants are given in the CAD program: $A = 1073\,\text{mm}^2$, $I_y = 3.00 \times 10^6\,\text{mm}^4$, $W_{y,el} = 3.93 \times 10^4\,\text{mm}^3$ and $z_{gc} = 72.1$ mm. Also, some measurements to check local buckling are needed: $b = 50$ mm, $t_f = 3.5$ mm, $h = 140$ mm, $t_w = 2$ mm, $t_w = 2$ mm, $z_{ue} = 77.5$ mm and, for the bottom flange, $t_b = 7$ mm and $h_n = 17$ mm (Figure 6.15).

Figure 6.15. Extruded aluminium profile (the third profile in Figure 6.7)

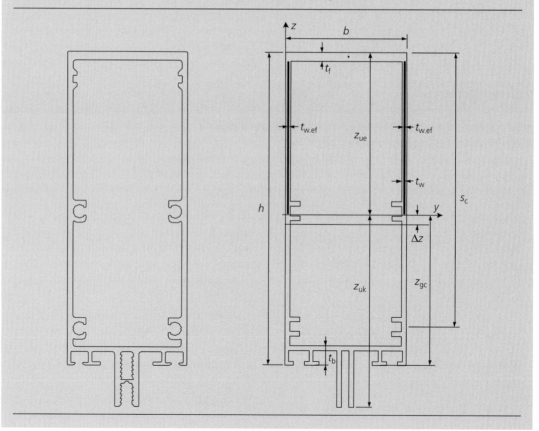

The influence of the web stiffeners (screw ports) close to the centre of the webs is small, and omitted when calculating the major axis moment resistance. For axial force, they may have noticeable influence: see Example 6.3.

Cross-section classification (*clause 6.1.4*)

Clause 6.1.4

In Table 6.4 (*Table 6.2*),

$$\varepsilon = \sqrt{250/160} = 1.25$$

Flange (*clause 6.1.4.3(1)*):

Clause 6.1.4.3(1)

$$\beta_f = (b - 2t_w)/t_f = (50 - 2 \times 2)/3.5 = 13.1$$

Limits in Table 6.4 (*Table 6.2*):

$$\beta_2 = 16\varepsilon = 16 \times 1.25 = 20$$

$$\beta_1 = 11\varepsilon = 11 \times 1.25 = 13.8$$

The flange is class 1.

Web (*clause 6.1.4.3(1)*) – stiffeners omitted: as the tension flange is much stiffened, the web is supposed to start at the middle of the bottom screw port. Then,

Clause 6.1.4.3(1)

$$\psi = -\frac{s_c - z_{ue}}{z_{ue} - t_f} = -\frac{125 - 77.5}{77.5 - 3.5} = -0.642$$

$$\beta_w = (0.7 + 0.3\psi)(s_c - t_f)/t_w = [(0.7 + 0.3 - 0.642)](125 - 3.5)/2 = 30.8$$

$$\beta_2 = 16\varepsilon = 16 \times 1.25 = 20$$

$$\beta_3 = 22\varepsilon = 22 \times 1.25 = 27.5$$

The web is class 4.

Shape factor

The section classification is class 4, and the shape factor is then based on the effective cross-section according to Table 6.8 (*Table 6.4* in *clause 6.2.5.1*). The compression flange is class 1, so only the webs need to be reduced.

Clause 6.2.5.1

For the material buckling class A according to *Table 3.2b*, the coefficients in *expression 6.12* is $C_1 = 32$ and $C_2 = 220$:

$$\rho_c = \min\left[C_1 \frac{\varepsilon}{\beta_w} - C_2\left(\frac{\varepsilon}{\beta_w}\right)^2, 1.0\right] = 32\frac{1.25}{30.8} - 220\left(\frac{1.25}{30.8}\right)^2 = 0.936 \qquad (6.12)$$

The effective thickness of the compression part of the webs is thus

$$t_{w,ef} = \rho_c t_w = 0.936 \times 2 = 1.872$$

where the width of the compression part of the web is

$$b_c = h - t_f - z_{gc} = 140 - 3.5 - 78.1 = 58.4\,\text{mm}$$

Again, the CAD program is used. The section moduli for the upper edge (ue) and bottom edge (be) of the section are found to be almost identical:

$$W_{ue} = 3.828 \times 10^4\,\text{mm}^3$$

$$W_{be} = 3.827 \times 10^4\,\text{mm}^3$$

Sometimes, an iteration procedure is needed to assure that the width of the compression part of the web coincides with the calculated neutral axis for the effective cross-section: see *clause 6.2.5.2(2)* and *clause 6.7.2(5)*. However, this is not necessary in this case as the effective cross-section is almost symmetric. The shape factor is then

$$\alpha_4 = \frac{\min(W_{\text{ue}}, W_{\text{be}})}{W_{\text{el}}} = \frac{3.827}{3.933} = 0.973$$

Bending moment resistance
The bending moment resistance is, from *expression 6.25*,

$$M_{\text{Rd}} = \alpha_4 W_{\text{el}} f_{\text{o}}/\gamma_{\text{M1}} = W_{\text{eff}} f_{\text{o}}/\gamma_{\text{M1}} = 3.827 \times 10^4 \times 160/1.1 = 5.57\,\text{kN}\,\text{m} \qquad (6.25)$$

Example 6.7: bending moment resistance of a welded member with a transverse weld

Two extruded channel sections are welded together to form a rectangular hollow section (Figure 6.16). Calculate the major axis bending moment resistance for

(a) the section without a transverse weld
(b) the section with a transverse butt weld across part of the web.

The material is EN AW-6082 T6, which, according to Table 3.1, belongs to buckling class A and has a proof strength $f_{\text{o}} = 260$ MPa. The partial factor of strength is $\gamma_{\text{M1}} = 1.1$.

The width $b = 100$ mm, the height $h = 300$ mm, the flange thickness $t_{\text{f}} = 10$ mm and the web thickness $t_{\text{w}} = 6$ mm.

(a) Resistance of the section without a transverse weld
Cross-section classification (*clause 6.1.4*). In Table 6.4 (*Table 6.2*),

$$\varepsilon = \sqrt{250/260} = 0.981$$

Flange (*clause 6.1.4.3(1)*):

$$\beta_{\text{f}} = (b - 2t_{\text{w}})/t_{\text{f}} = (160 - 2 \times 6)/10 = 14.8$$

Figure 6.16. Welded aluminium profile

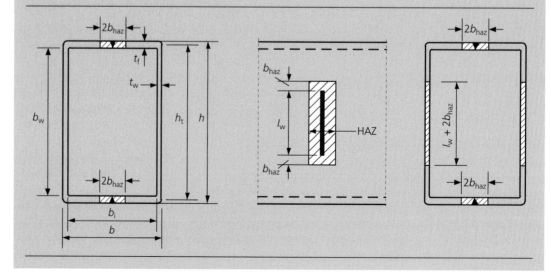

Clause 6.2.5.2(2)
Clause 6.7.2(5)

Clause 6.1.4
Clause 6.1.4.4

Clause 6.1.4.3(1)

Limits in Table 6.4 (*Table 6.2*) for buckling class A, **with welds**:

$$\beta_2 = 13\varepsilon = 13 \times 0.951 = 12.7$$

$$\beta_3 = 18\varepsilon = 18 \times 0.981 = 17.7$$

The flange is class 3.

Web (*clause 6.1.4.3(1)*):

Clause 6.1.4.3(1)

$$\beta_w = 0.4h_w/t_w = 0.4(300 - 20)/6 = 18.7$$

Limits in Table 6.4 (*Table 6.2*) for buckling class A, **without welds**:

$$\beta_2 = 16\varepsilon = 16 \times 0.981 = 15.7$$

$$\beta_3 = 22\varepsilon = 22 \times 0.981 = 21.6$$

The web is class 3.

HAZs (*clause 6.1.6*). The reduction factor for the strength in the HAZ is found in *Table 3.2b*, and the extent is found in *clause 6.1.6.3*:

Clause 6.1.6
Clause 6.1.6.3

$$\rho_{o,haz} = 0.48, \ b_{haz} = 30 \text{ mm}$$

for $t_f = 10$ mm

The effective thickness within the HAZ will be

$$t_{haz} = \rho_{o,haz}t_f = 0.48 \times 10 = 4.8 \text{ mm}$$

The elastic section modulus allowing for the HAZ is found by deleting the difference between the flange thickness and the effective thickness within the width $2b_{haz}$ from the gross cross-section:

$$I_y = \frac{1}{12}\left(bh^3 - (b - 2t_w)(h - 2t_f)^3\right)$$
$$= \frac{1}{12}\left(16 \times 300^3 - 148 \times 280^3\right) = 8.926 \times 10^7 \text{ mm}^4$$

$$W_{el} = I_y 2/h = 8.926 \times 10^7 \times 2/300 = 5.95 \times 10^5 \text{ mm}^3$$

$$I_{y,haz} = I_y - 2b_{haz}(t_f - t_{haz})2\left(\frac{h_t}{2}\right)^2$$
$$= 8.926 \times 10^7 - 2 \times 30(10 - 4.80)2 \times 145^2 = 7.61 \times 10^7 \text{ mm}^4$$

$$W_{el,haz} = I_{y,haz}2/h = 7.61 \times 10^7 \times 2/300 = 5.08 \times 10^5 \text{ mm}^3$$

The plastic section modulus allowing for the HAZ is

$$W_{pl,haz} = \tfrac{1}{4}\left(bh^2 - (b - 2t_w)(h - 2t_f)^2\right) - 2b_{haz}(t_f - t_{haz})h_t$$
$$= \tfrac{1}{4}\left(160 \times 300^2 - 148 \times 280^2\right) - 2 \times 30(10 - 4.80)290 = 6.09 \times 10^5 \text{mm}^4$$

Shape factor (*clause 6.2.5.1*). For cross-section class 3 the shape factor $= 1.0$, or may alternatively be calculated using *expression 6.27* in *clause 6.2.5.1*. As the web area is a

Clause 6.2.5.1
Clause 6.2.5.1

large part of the cross-section, *expression 6.27* is used:

$$\alpha_{3,w} = \frac{W_{el,haz}}{W_{el}} + \left(\frac{\beta_3 - \beta}{\beta_3 - \beta_2}\right)\left(\frac{W_{pl,haz} - W_{el,haz}}{W_{el}}\right) \quad (6.27)$$

where the cross-section part with the smallest value of the ratio $(\beta_3 - \beta)/(\beta_3 - \beta_2)$ is the critical part. For the flange and the web,

$$(\beta_{3f} - \beta_f)/(\beta_{3f} - \beta_{2f}) = (17.7 - 14.8)/(17.7 - 12.7) = 0.581$$

$$(\beta_{3w} - \beta_w)/(\beta_{3w} - \beta_{2w}) = (21.6 - 18.7)/(21.6 - 15.7) = 0.494$$

The web is decisive:

$$\alpha_{3,u} = \frac{W_{el,haz}}{W_{el}} + \left(\frac{\beta_3 - \beta}{\beta_3 - \beta_2}\right)\left(\frac{W_{pl,haz} - W_{el,haz}}{W_{el}}\right)$$

$$= \frac{5.08}{5.95} + 0.494\frac{6.09 - 5.08}{5.95} = 0.937 \quad (6.27)$$

Bending moment resistance. The bending moment resistance is, from *expression 6.25*,

$$M_{Rd} = \alpha_{3,w}W_{el}f_o/\gamma_{M1} = 0.937 \times 5.95 \times 10^5 \times 260/1.1 = 132\,\text{kN m} \quad (6.25)$$

Clause 6.2.5.1

(b) Resistance of the section with a transverse weld (*clause 6.2.5.1*)
The resistance in section with the transverse weld is given by *expression 6.24b*:

$$M_{u,Rd} = W_{u,eff,haz}f_u/\gamma_{M2} \quad (6.24b)$$

where $W_{u,eff,haz}$ is the effective section modulus, obtained using a reduced thickness $\rho_c t$ for class 4 parts and a reduced thickness $\rho_{u,haz}t$ for the HAZ material, whichever is smaller.

The cross-section classification and the effective thickness are the same as for the section without a transverse weld (class 3), which is why there is no reduction due to local buckling.

The reduction factor for the ultimate strength in the HAZ is, from Table 3.1 for EN AW-6082-T6, $\rho_{u,haz} = 0.60$. Thus, the section modulus with an allowance for the HAZ due to a longitudinal weld and a localised transverse weld is

$$I_{u,eff,haz} = I_y - 2b_{haz}t_f(1 - \rho_{u,haz})2\left(\frac{h_t}{2}\right)^2 - (1 - \rho_{u,haz})2t_w(l_w + 2b_{haz})^3/12$$

$$= 8.926 \times 10^7 - 2 \times 30 \times 10(1 - 0.6) \times 2 \times 145^2 - (1 - 0.6)2 \times 6(120 + 2 \times 30)^3/12$$

$$= 8.592 \times 10^7\,\text{mm}^4$$

$$W_{u,eff,haz} = \frac{I_{u,eff,haz} \times 2}{h} = \frac{8.59 \times 10^7 \times 2}{300} = 5.73 \times 10^5\,\text{mm}^3$$

Bending moment resistance at the section with a transverse weld. The bending moment resistance is, according to *expression 6.24b*,

$$M_{u,Rd} = W_{u,eff,haz}f_u/\gamma_{M2} = 5.73 \times 10^5 \times 310/1.25 = 142\,\text{kN m} \quad (6.24b)$$

This is actually larger than the resistance of the member with longitudinal welds only, which is $M_{c,Rd} = 132\,\text{kN m}$. So, in this case the HAZ in the transverse welds does not reduce the bending moment resistance of the member.

The strength in the weld according to Table 8.8 for filler metal 5356 is $f_w = 210\,\text{MPa}$, which is greater than the strength in the HAZ, and thus not critical.

Other examples

Other examples where the resistance of cross-sections in bending is included are Examples 6.10, 6.11, 6.12, 6.13 and 6.14.

For combined bending and axial force, the designer should refer to *clause 6.2.9*.

Clause 6.2.9

In the compression zone of a cross-section in bending (as for cross-sections under uniform compression), no allowance needs to be made for fastener holes (where fasteners are present) except for oversized holes and slotted holes. Fastener holes in the tension flange and the tensile zone of the web should be checked for net section resistance according to *clause 6.2.5.1(2)*.

Clause 6.2.5.1(2)

6.2.6 Shear

The resistance of cross-sections to shear is covered in *clause 6.2.6*. The design shear force is denoted by V_{Ed} (the shear force design effect). The design shear resistance of a cross-section is denoted by V_{Rd}, and may be calculated based on an elastic distribution with a moderate allowance for plastic redistribution of shear stress. The shear stress distribution in a rectangular section and in an I section, based on purely elastic behaviour, is shown in Figure 6.17.

Clause 6.2.6

In both cases in Figure 6.17, the shear stress varies parabolically with depth, with the maximum value occurring at the neutral axis. However, for the I section, the difference between the maximum and minimum values for the web, which carries almost all the vertical shear force, is relatively small. Consequently, by allowing a degree of plastic redistribution of the shear stress, design can be simplified to working with the average shear stress, defined as the total shear force V_{Ed} divided by the area of the web (or the equivalent shear area A_v).

Since the yield stress in shear is approximately $1/\sqrt{3}$ of its yield stress in tension, *clause 6.2.6(2)* therefore defines the shear resistance as

Clause 6.2.6(2)

$$V_{Rd} = A_v \frac{f_o}{\sqrt{3}\gamma_{M1}} \qquad (6.29)$$

The shear area A_v is the area of the cross-section that can be mobilised to resist the applied shear force with a moderate allowance for plastic redistribution, and, for sections where the load is applied parallel to the web, this is essentially the area of the web. Expressions for the determination of the shear area A_v for structural aluminium cross-sections are given in *clause 6.2.6(3)*, and are repeated below.

Clause 6.2.6(3)

For non-slender sections ($h_w/t_w < 39\varepsilon$) containing shear webs,

$$A_v = \sum_{i=1}^{n} \left[(h_w - \sum d)(t_w)_i - (1 - \rho_{o,haz})b_{haz}(t_w)_i \right] \qquad (6.30)$$

Figure 6.17. Distribution of shear stress in rectangular and I cross-sections

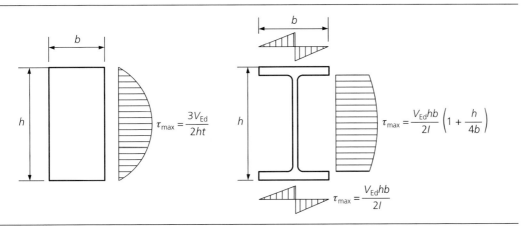

where:

h_w is the depth of the web between flanges

b_{haz} is the total depth of HAZ material occurring between the clear depth of the web between flanges (for sections with no welds, $\rho_{o,haz} = 1$; if the HAZ extends the entire depth of the web panel, then $b_{haz} = h_w - \sum d$)

t_w is the web thickness

d is the diameter of holes along the shear plane

n is the number of webs.

For a solid bar,

$$A_v = 0.8 A_e$$

For a round tube

$$A_v = 0.6 A_e$$

where A_e is the full section area of a non-welded section, and the effective section area obtained by taking a reduced thickness $\rho_{o,haz} t$ for the HAZ material of a welded section.

Clauses 6.7.4–6.7.6   For slender webs and stiffened webs, reference should be made to **clauses 6.7.4 to 6.7.6.**

6.2.7   Torsion

Clause 6.2.7   The resistance of cross-sections to torsion is covered in **clause 6.2.7.** Torsional loading can arise in two ways: either due to an applied torque (pure twisting) or due to a transverse load applied eccentrically to the shear centre of the cross-section (twisting plus bending). In engineering

Clause 6.2.7   structures it is the latter that is the most common, and pure twisting is relatively unusual. Con-

Clause 6.2.8   sequently, **clauses 6.2.7, 6.2.8** and **6.2.10** provide guidance for torsion acting in combination with

Clause 6.2.10   other effects (bending, shear and axial force).

There are many means to avoid torsion due to an applied load. In Figure 6.18(a) the torsion moment is Fe. In Figure 6.18(b), a stiffener is added such that loading can act through the shear centre SC, and in Figure 6.18(c) this is achieved by the shape of the cross-section. The torsional stiffness of the hollow section in Figure 6.18(d) is many hundreds of times larger than the open section in Figure 6.18(e). Lateral bracing can also be added to the two flanges (Figure 6.18(f)), or rotation prevented by fixing the slab to the flange (Figure 6.18(g)).

The torsional moment design effect T_{Ed} is made up of two components: the Saint Venant torsion $T_{t,Ed}$ and the warping torsion $T_{w,Ed}$.

Saint Venant torsion is the uniform torsion that exists when the rate of change of the angle of twist along the length of a member is constant. In such cases, the longitudinal warping deformations (which accompany twisting) are also constant, and the applied torque is resisted by a single set of shear stresses, distributed around the cross-section.

Figure 6.18. Torsion moment and means to avoid torsion

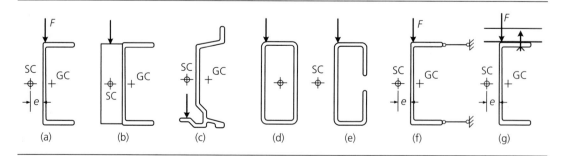

Warping torsion exists where the rate of change of the angle of twist along the length of a member is not constant; in which case, the member is said to be in a state of non-uniform torsion (Vlasov torsion). Such non-uniform torsion may occur either as a result of non-uniform loading (i.e. varying torque along the length of the member) or due to the presence of longitudinal restraint to the warping deformations. For non-uniform torsion, longitudinal direct stresses and an additional set of shear stresses arise. Therefore, as noted in **clause 6.2.7.2(3)**, there are three sets of stresses that should be considered:

Clause 6.2.7.2(3)

- shear stresses $\tau_{t,Ed}$ due to the Saint Venant torsion $T_{t,Ed}$
- shear stresses $\tau_{w,Ed}$ due to the warping torsion $T_{w,Ed}$
- longitudinal direct stresses $\sigma_{w,Ed}$ due to the warping (from the bimoment B_{Ed}).

Depending on the cross-section classification, torsional resistance may be verified plastically with reference to **clause 6.2.7**, or elastically by adopting the yield criterion of *expression 6.15* (see **clause 6.2.1(5)**).

Clause 6.2.7
Clause 6.2.1(5)

Clause 6.2.7.2(6) allows useful simplifications for the design of torsion members. For closed-section members (such as cylindrical and rectangular hollow sections), for which the torsional rigidities are very large, Saint Venant torsion dominates, and warping torsion may he neglected. Conversely, for open sections, such as I or H sections, for which the torsional rigidities are low, Saint Venant torsion may be neglected.

Clause 6.2.7.2(6)

Remember that if the resultant force is acting through the shear centre, there is no torsional moment due to that loading. Formulae for the shear centre for some common cross-sections are given in *Annex J*.

For the case of combined shear force and torsional moment, **clause 6.2.7.3** defines a reduced plastic shear resistance $V_{T,Rd}$ that must be demonstrated to be greater than the design shear force V_{Ed}. $V_{T,Rd}$ may be derived from *expressions 6.35, 6.36* and *6.37* (not repeated here).

Clause 6.2.7.3

6.2.8 Bending and shear

Bending moments and shear forces acting in combination on structural members are common. However, in the majority of cases the effect of shear force on the moment resistance is negligible, and may be ignored. **Clause 6.2.8(2)** states that, provided the applied shear force is less than half of the plastic shear resistance of the cross-section, its effect on the moment resistance may be neglected. The exception to this is where shear buckling reduces the resistance of the cross-section, as described in Section 6.7.6 of this guide.

Clause 6.2.8(2)

For cases where the applied shear force is greater than half of the plastic shear resistance of the cross-section, the moment resistance should be calculated using a reduced design strength for the shear area, given by *expression 6.38*:

$$f_{o,V} = f_o(1 - (2V_{Ed}/V_{Rd} - 1)^2) \tag{6.38}$$

where V_{Rd} is obtained from Section 6.2.6. If torsion is present, V_{Rd} in *expression 6.38* is replaced by $V_{T,Rd}$ (see Section 6.2.7), but $f_{o,V} = f_o$ for $V_{Ed} \leq 0.5V_{T,Rd}$.

In the case of an equal-flanged I section classified as class 1 or 2 in bending, the resulting value of the reduced moment resistance $M_{v,Rd}$ is

$$M_{v,Rd} = t_f b_f(h - t_f)\frac{f_o}{\gamma_{M1}} + \frac{t_w h_w^2}{4}\frac{f_{o,V}}{\gamma_{M1}} \tag{6.39}$$

where h is the total depth of the section and h_w is the web depth between inside flanges.

In the case of an equal-flanged I section classified as class 3 in bending, the resulting value of $M_{v,Rd}$ is given by *expression 6.39*, but with the denominator 4 in the second term replaced by 6.

For sections classified as class 4 in bending or affected by HAZ softening, see **clause 6.7.6**.

Clause 6.7.6

Clause 6.7.6

For the interaction of bending, shear force and transverse loads, the rules for plate girders in *clause 6.7.6* should be used.

An example of the application of the cross-section rules for combined bending and shear force is given in Example 6.8.

Example 6.8: cross-section resistance under combined bending and shear

An extruded profile is to be used as a short-span ($L = 1.2$ m), simply supported, laterally restrained beam. It is to be designed for a central concentrated load of $F_{Ed} = 180$ kN, as shown in Figure 6.19. The arrangement results in a maximum shear force $V_{Ed} = 90$ kN and a maximum bending moment $M_{Ed} = 54$ kN m.

Check the resistance of the beam if made of EN AW-6082 T6.

Section properties

Section height	$h = 220$ mm
Flange width	$b = 100$ mm
Flange thickness	$t_f = 8$ mm
Web thickness	$t_w = 6$ mm
Fillet radius	$r = 12$ mm
Web height	$b_w = h - 2t_f - 2r = 154$ mm
EN AW-6082 T6	$f_o = 260$ MPa
Partial safety factor	$\gamma_{M1} = 1.1$

Clause 6.1.4

Cross-section classification (*clause 6.1.4*)

$$\varepsilon = \sqrt{250/f_o} = \sqrt{250/260} = 0.981$$

Outstand flanges (*expression 6.1*):

$$\beta_f = (b - t_w - 2r)/2t_f = (100 - 6 - 2 \times 14)/(2 \times 9) = 3.67$$

Limits for classes 1 and 2:

$$\beta_1 = 3\varepsilon = 2.94 < \beta_f$$

$$\beta_2 = 4.5\varepsilon = 4.41 > \beta_f$$

The flange is class 2.

Web – internal part with stress gradient $\psi = -1$ (*expression 6.1*):

$$\beta_w = 0.4b_w/t_w = 0.4 \times 180/6 = 12$$

Figure 6.19. General arrangement – loading and cross-section notation

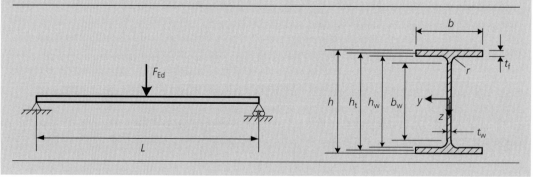

Limits for classes 1 and 2:

$$\beta_1 = 11\varepsilon = 10.8 < \beta_w$$

$$\beta_2 = 16\varepsilon = 15.7 > \beta_w$$

The web is class 2.

The overall cross-section classification is class 2. The resistance is therefore the plastic bending moment resistance.

Bending moment resistance of the cross-section (*clause 6.2.5*)

Clause 6.2.5

Including the fillets and using the notation (see Figure 6.19), $h_t = h - t_f = 220 - 8 = 219$ mm and $h_w = h - 2t_f = 220 - 2 \times 8 = 204$ mm, we have

$$W_{pl} = bt_f h_t + \tfrac{1}{4}t_w h_w^2 + 2r^2(h_w - r) - \pi r^2 \left[h_w - 2r\left(1 - \frac{4}{3\pi}\right) \right]\tfrac{1}{2}$$

$$W_{pl} = 100 \times 8 \times 212 + \frac{1}{4}(6 \times 204^2) + 2 \times 12^2(204 - 12) - \pi$$

$$\times 12^2 \left[204 - 2 \times 12\left(1 - \frac{4}{3\pi}\right) \right]\tfrac{1}{2}$$

$$= 2.443 \times 10^5 \, \text{mm}^3$$

and the bending moment resistance is

$$M_{Rd} = f_o W_{pl}/\gamma_{M1} = 260 \times 2.443 \times 10^5/1.1 = 57.7 \text{ kN m}$$

which is larger than $M_{Ed} = 54$ kN m.

Shear resistance of the cross-section (*clause 6.2.6*)

Clause 6.2.6

$h_w/t_w = 204/6 = 34 < 39\varepsilon = 38.2$, which means that shear buckling need not be checked.

$$A_v = h_w t_w = 204 \times 6 = 1224 \, \text{mm}^2$$

$$V_{Rd} = A_v f_o/\sqrt{3}\gamma_{M1} = 1224 \times 260/(\sqrt{3} \times 1.1) = 167 \text{ kN}$$

which is larger than $V_{Ed} = 90$ kN, but $V_{Ed} > V_{Rd}/2$, so combined bending and shear needs to be checked.

Combined bending and shear (*clause 6.2.8*)

Clause 6.2.8
Clause 6.2.8(3)

The reduced moment resistance is found in *clause 6.2.8(3)*:

$$f_{oV} = f_o\left[1 - \left(\frac{2V_{Ed}}{V_{Rd}} - 1\right)^2\right] = 260\left[1 - \left(\frac{2 \times 90}{167}\right)^2\right] = 258 \text{ MPa} \qquad (6.38)$$

$$M_{v,Rd} = t_f b_f (h - t_f)\frac{f_o}{\gamma_{M1}} + \frac{t_w h_w^2}{4}\frac{f_{o,V}}{\gamma_{M1}} = 8 \times 100 \times 212\frac{260}{1.1} + \frac{6 \times 204^2}{4}\frac{258}{1.1}$$

$$= 54.8 \text{ kN m} \qquad (6.39)$$

$$M_{v,Rd} = 54.8 \text{ kN m} > M_{Ed} = 54 \text{ kN m}$$

Cross-section resistance to combined bending and shear is acceptable.

6.2.9 Bending and axial force

The design of cross-sections subjected to combined bending and axial force is described in *clause 6.2.9*. Bending may be about one or both principal axes, and the axial force may be tensile or compressive. The interaction formulae are valid for all four cross-section classes; however, two sets of formulae are given for open cross-sections and closed cross-sections.

In the following, the strength and behaviour of beam column segments subjected to compression combined with biaxial bending are presented. As the name implies, here we are only concerned with short beam columns for which the effect of lateral deflections on the magnitudes of bending moments is negligible. As a result, the maximum strength occurs when the entire cross-section is fully plastic or yielded in the case of elastic–plastic material (e.g. mild steel) or when the maximum strain (or stress) attains some prescribed value in the case of a hardening material such as aluminium. This is a stress problem of the first order. Material yielding or failure is the primary cause of the strength limit of the member. The method of analysis in predicting this limiting strength is presented here as a background to Eurocode 9.

Rectangular section – plastic theory

If the slenderness b/t of the cross-sectional parts is small and the strain at rupture large, the whole cross-section may yield. For a solid, rectangular section of ideal elastic–plastic material, the relation between the axial force and bending moments for the rectangular stress distribution is easily derived.

The stress distribution according to Figure 6.20 corresponds to the section resultants

$$N = f_y b(h - 2z) \tag{D6.6}$$

$$M = f_y bz(h - z) \tag{D6.7}$$

If z is eliminated and the following notation for the plastic compression force and moment is introduced,

$$N_{pl} = f_y bh \tag{D6.8}$$

$$M_{pl} = \frac{f_y bh^2}{4} \tag{D6.9}$$

then, after rearranging the expressions, we arrive at the interaction formula:

$$\left(\frac{N}{N_{pl}}\right)^2 + \frac{M}{M_{pl}} = 1 \tag{D6.10}$$

This equation represents a parabola, as shown in Figure 6.20.

Figure 6.20. Rectangular cross-section subject to an axial force and a bending moment

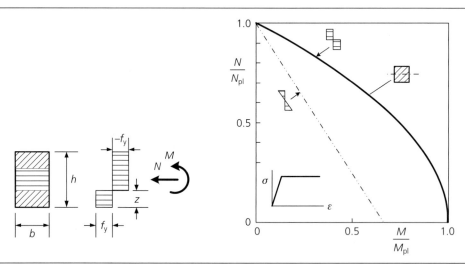

Figure 6.21. Stress–strain relationship according to Ramberg–Osgood and interaction curves for rectangular cross-section (ε_u in %)

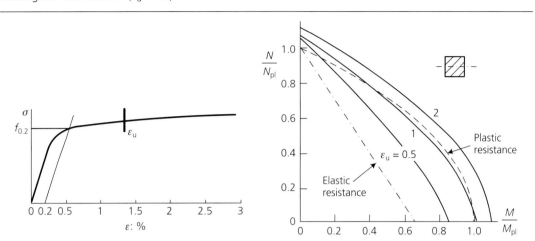

Rectangular section – strain hardening material

Similar curves can be derived for a strain-hardening material as aluminium. The bending moment is given by integrating the stresses over the cross-section, assuming a linear strain distribution. As the strain ε in the Ramberg–Osgood expression (see *Annex E* and Section 9.5 of this guide) is a function of the stress σ, then numerical integration is used, dividing the cross-section into small elements. The shape of the curve depends on the stress–strain relationship and the limiting strain ε_y. Curves are shown in Figure 6.21 for aluminium having a strain-hardening parameter $n = 15$, in the Ramberg–Osgood expression

$$\varepsilon = \frac{\sigma}{E} + 0.002\left(\frac{\sigma}{f_{0,2}}\right)^n \tag{E.12}$$

The stress–strain relationship of this material is shown in Figure 6.21.

If the limiting compressive strain is $\varepsilon_u = 0.01$ (1%), which is about twice the strain corresponding to the proof stress $f_{0.2}$ ($= f_o$), then the curve is very close to that for an ideal elastic–plastic material (dashed curve).

I section – strain-hardening material

Interaction curves are given in Figure 6.22 for aluminium beam column segments with I cross-sections. Especially for minor axis bending, the curves are strongly convex upwards.

Figure 6.22. Interaction curves for I cross-sections (ε_u in %): (a) y axis bending; (b) z axis bending

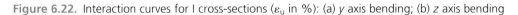

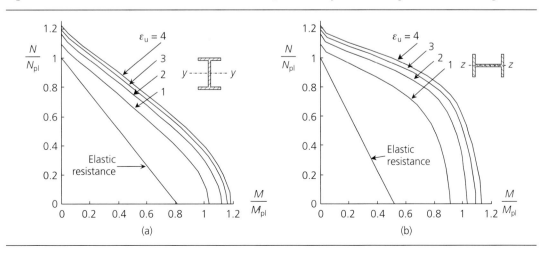

As noted above, the curves for $\varepsilon_u = 0.01$ are very close to those for an ideal elastic–plastic material. Then, the following curves and expressions for an elastic–plastic material are approximately valid for aluminium.

I section – interaction formulae

The interaction formula for rectangular cross-sections of an ideal elastic–plastic material was easy to derive. For other cross-sections the derivation can, in principle, be done in the same way, but the expressions often become too complicated for design. For I sections, the interaction formulae will, after some simplification, be:

■ for y-axis bending:

$$\frac{N}{N_{pl}} + \frac{M}{M_{pl,y}}(1 - 0.5\alpha_w) = 1 \qquad \text{but } M_{N,y} \le M_{pl,y} \tag{D6.11}$$

■ for z-axis bending:

$$\left(\frac{N/N_{pl} - \alpha_w}{1 - \alpha_w}\right)^2 + \frac{M}{M_{pl,z}} = 1 \qquad \text{for } N/N_{pl} > \alpha_w \quad \text{but } M \le M_{pl,z} \tag{D6.12}$$

where $\alpha_w = (A - 2bt_f)/A$, but $\alpha_w \le 0.5$.

For y axis bending, the formula corresponds to curve 3 in Figure 6.23, and for z-axis bending, curve 1. For z axis bending, the flanges carry the bending moment, and as long as the axial force is less than what the web can resist, the moment resistance remains unaffected. The curve for the axial force and z axis bending is therefore approximately equal to the curve for a rectangular cross-section uplifted for a large moment.

Yielding of class 3 web is limited by local buckling. At the limit between class 3 and class 4, the resistance is determined by the fact that the yield strength is reached in the extreme fibre of the beam, and the stress distribution is therefore linear according to elastic theory (the class 4 cross-section is based on the effective cross-section). The interaction curves will then also be straight lines, which, for $N = 0$, start at the moment resistance $M_{el} = W_{el}f_y$; that is, for M_{el}/M_{pl} on the abscissa according to curve 5 in Figure 6.23.

Figure 6.23. Interaction curves for axial force and bending moment for beams of rectangular cross-section or I section according to plastic theory (curves 1, 2 and 3) and elastic theory (curves 4 and 5)

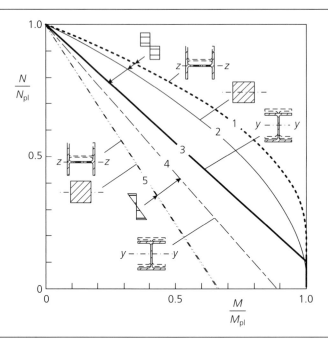

Figure 6.24. The moment resistance as a function of the web slenderness

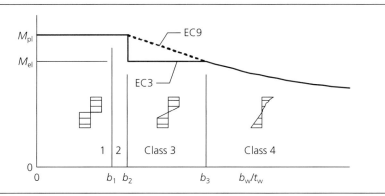

In fact, these straight lines only apply for slenderness on the limit between class 3 and class 4 cross-sections. In class 3 the beam may yield to a certain degree, and for slenderness close to the limit of class 2, the whole web will yield. See Figure 6.24. For simplification, this is not used in EN 1993-1-1 (steel), but the resistance is $M_{el} = W_{el}f_y$ within the whole class 3 cross-section (with some exceptions). In EN 1999-1-1 (aluminium) there is linear interpolation between $M_{el} = W_{el}f_y$ and $M_{pl} = W_{pl}f_y$, corresponding to the dotted line in Figure 6.24.

For axial force and bending, the difference between the Eurocodes for steel and aluminium are even more pronounced because the interaction formulae are different. The formulae according to EN 1999-1-1 are the following (***clause 6.2.9.1***):

Clause 6.2.9.1

$$\left(\frac{N_{Ed}}{N_{Rd}}\right)^{\xi_0} + \frac{M_{y,Ed}}{M_{y,Rd}} \le 1.00 \tag{6.40}$$

$$\left(\frac{N_{Ed}}{N_{Rd}}\right)^{\eta_0} + \left(\frac{M_{y,Ed}}{M_{y,Rd}}\right)^{\gamma_0} + \left(\frac{M_{z,Ed}}{M_{z,Rd}}\right)^{\xi_0} \le 1.00 \tag{6.41}$$

where:

$\eta_0 = 1.0$ or may alternatively be taken as

$\alpha_z^2\alpha_y^2 \quad$ but $1 \le \eta_0 \le 2$ $\tag{6.42a}$

$\gamma_0 = 1.0$ or may alternatively be taken as

$\alpha_z^2 \quad$ but $1 \le \gamma_0 \le 1.56$ $\tag{6.42b}$

$\xi_0 = 1.0$ or may alternatively be taken as

$\alpha_y^2 \quad$ but $1 \le \xi_0 \le 1.56$ $\tag{6.42c}$

Clause 6.2.3

N_{Rd} is the axial force resistance according to ***clause 6.2.3*** or ***6.2.4***, respectively

Clause 6.2.4

$M_{y,Rd}$, $M_{z,Rd}$ are the bending moment resistances with respect to the y–y and z–z axes according to ***clause 6.2.5***

Clause 6.2.5

α_y, α_z are the shape factors for bending about the y and z axes (see ***clause 6.2.5***).

Clause 6.2.5

Interaction curves corresponding to *expressions 6.40* and *6.41* for $M_{y,Ed} = 0$ are given in Figure 6.25. The series of curves for class 3 cross-sections are the result of the exponents, which are functions of the shape factors α_y and α_z. For y axis bending, the influence of the cross-section slenderness is not very pronounced since α_y does not vary much, usually between 1.0 and about 1.15. For z axis bending, however, the resistance may be doubled using the Eurocode 9 interaction formulae instead of using the elastic moment resistance as in Eurocode 3 for class 3 cross-sections.

Figure 6.25. Interaction curves for I beams in y axis and z axis bending and axial compression or tension

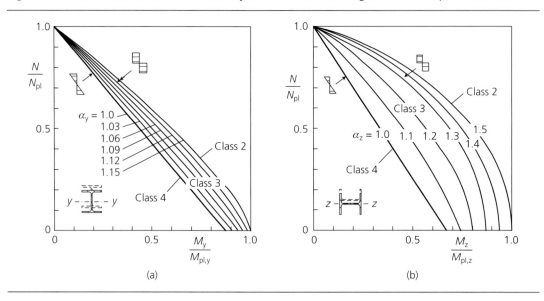

(a)

(b)

Hollow sections and solid cross-sections

Clause 6.2.9.2

In *clause 6.2.9.2* the following interaction formula is given for hollow sections and solid cross-sections:

$$\left(\frac{N_{\mathrm{Ed}}}{N_{\mathrm{Rd}}}\right)^{\psi} + \left[\left(\frac{M_{\mathrm{y,Ed}}}{M_{\mathrm{y,Rd}}}\right)^{1.7} + \left(\frac{M_{\mathrm{z,Ed}}}{M_{\mathrm{z,Rd}}}\right)^{1.7}\right]^{0.6} \leq 1.00 \qquad (6.43)$$

where, for hollow sections, $\psi = 1.3$ for class 1 and class 2 cross-sections and $\psi = 1.0$ for class 3 and class 4 cross-sections, or, alternatively, for all classes of cross-sections, ψ may be taken as $\alpha_y \alpha_z$, but $1 \leq \psi \leq 1.3$. For solid sections, $\psi = 2$.

For uniaxial bending, say $M_{\mathrm{z,Rd}} = 0$, the exponent will be $1.7 \times 0.6 \approx 1.0$. For a massive rectangular cross-section with $\psi = 2$ this means that the result is the same as in expression D6.10, as expected. For a rectangular class 1 or 2 hollow section, the exponent ψ will be $\psi = 1.3$, and the result is similar as with *expression 6.40*.

Bi-axial bending and compression

Figure 6.26 illustrates the result for biaxial bending of a rectangular section and for a class 2 H

Figure 6.26. Interaction diagrams for rectangular and class 2 H sections in bi-axial bending and compression

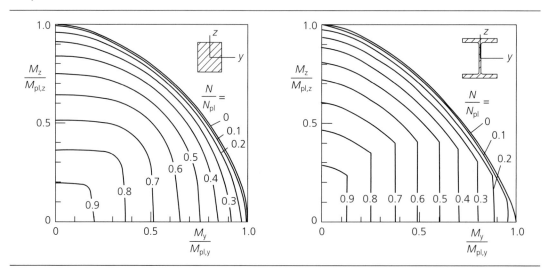

Figure 6.27. Interaction diagrams for $N/N_{pl} = 0.5$ and 0.75 for class 3 H sections in bi-axial bending and compression for shape factors varying from 1.00 to W_{pl}/W_{el}

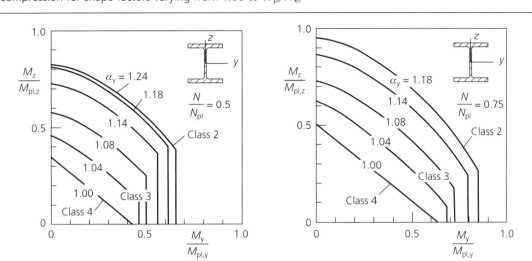

section. For class 4 cross-sections, plastic strain cannot occur, so the interaction curves will be straight lines according to the lines $\alpha_y = 1.00$ in Figure 6.27.

For H cross-sections in class 3, there is a gradual change from the curves in Figure 6.26 to straight lines as illustrated in Figure 6.27. This is achieved by the exponents, which are functions of the shape factors. Note that the moments on the axes are divided by the plastic moment resistance in all diagrams.

If the interpolation formulae (*expressions 6.26* or *6.27*) are not utilised, the resistance for class 3 cross-sections will correspond to the straight lines $\alpha_y = 1.00$, sometimes losing more than half of their strength.

Example 6.9: cross-section resistance of a square hollow section under combined bending and compression

A member is to be designed to carry a combined bending moment $M_{Ed} = 8$ kN m and an axial force $N_{Ed} = 240$ kN. In this example, a cross-sectional check is performed on a square hollow extrusion made of aluminium EN AW-6082 T6 (Figure 6.28).

Section properties and material

Section width	$b = 100$ mm
Section thickness	$t = 5$ mm
Inner dimension	$b_i = b - 2t = 90$ mm
EN AW-6082 T6	$f_o = 260$ MPa
Partial safety factor	$\gamma_{M1} = 1.1$

Figure 6.28. Square cross-section

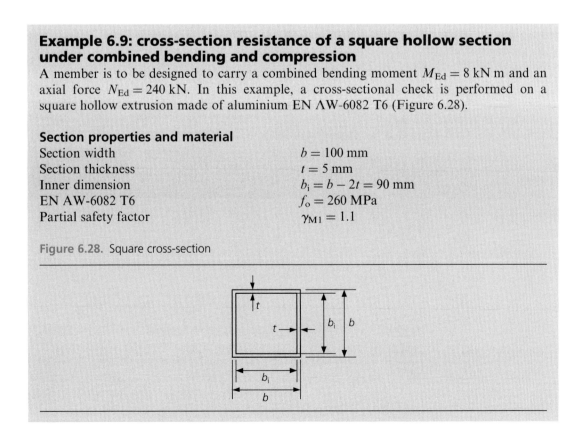

Clause 6.1.4

Cross-section classification (*clause 6.1.4*)

$$\varepsilon = \sqrt{250/f_o} = \sqrt{250/260} = 0.981$$

$$\beta = (b - 2t)/t = 90/5 = 18$$

Limits for classes 2 and 3:

$$\beta_2 = 16\varepsilon = 15.7 < \beta$$

$$\beta_3 = 22\varepsilon = 21.6 > \beta$$

The cross-section classification is class 3 for both compression and bending.

Clause 6.2.5

Bending moment resistance of the cross-section (*clause 6.2.5*)

$$I = \frac{1}{12}(b^4 - b_i^4) = \frac{1}{12}(100^4 - 90^4) = 2.87 \times 10^6 \, \text{mm}^4$$

$$W_{el} = 2I/b = 2 \times 2.87 \times 10^6/100 = 57\,320 \, \text{mm}^3$$

$$W_{pl} = \tfrac{1}{4}(b^3 - b_i^3) = \tfrac{1}{4}(100^3 - 90^3) = 67\,750 \, \text{mm}^3$$

The shape factor is

$$\alpha = 1 + \frac{\beta_3 - \beta}{\beta_3 - \beta_2}\left(\frac{W_{pl}}{W_{el}} - 1\right) = 1 + \frac{21.6 - 18}{21.6 - 15.7}\left(\frac{67\,750}{57\,320} - 1\right) = 1.11$$

and the bending resistance is

$$M_{Rd} = \alpha f_o W_{el}/\gamma_{M1} = 1.11 \times 260 \times 57\,320/1.1 = 15.0 \, \text{kN m}$$

Clause 6.2.4

Axial force resistance of the cross-section (*clause 6.2.4*)
The class 3 cross-section means that there is no reduction due to local buckling:

$$A = (b^2 - b_i^2) = (100^2 - 90^2) = 1900 \, \text{mm}^2$$
$$N_{Rd} = f_o A/\gamma_{M1} = 260 \times 1900/1.1 = 449 \, \text{kN}$$

Clause 6.2.9.2

Interaction (*clause 6.2.9.2*)
In *expression 6.43*, $M_{z,Ed} = 0$, so the expression is simplified to

$$\left(\frac{N_{Ed}}{N_{Rd}}\right)^{\psi} + \frac{M_{y,Ed}}{M_{y,Rd}} \leq 1.00 \qquad\qquad (6.43)$$

where

$$\psi = \alpha_y \alpha_z = 1.11 \times 1.11 = 1.23 < 1.3$$

$$M_{y,Ed} = M_{Ed} = 8 \, \text{kN m}$$

$$M_{y,Rd} = M_{Rd} = 15.0 \, \text{kN m}$$

$$\left(\frac{N_{Ed}}{N_{Rd}}\right)^{\psi} + \frac{M_{y,Ed}}{M_{y,Rd}} = \left(\frac{240}{449}\right)^{1.23} + \frac{8}{15.0} = 0.993 < 1.00$$

The cross-section resistance to combined bending and compression is acceptable.

6.2.10 Bending, shear and axial force

Where shear and axial force are present, allowance should be made for the effect of both the shear force and axial force on the resistance of the moment (*clause 6.2.10*). Provided that the design value of the shear force V_{Ed} does not exceed 50% of the shear resistance V_{Rd}, no reduction in the resistances defined for bending and the axial force in Section 6.2.9 need be made, except where shear buckling reduces the section resistance (see *clause 6.7.6*). *Clause 6.2.10*

Clause 6.7.6

Where V_{Ed} exceeds 50% of V_{Rd}, the design resistance of the cross-section to combinations of moments and axial force should be reduced using the reduced yield strength according to *expressions 6.46* and *6.47*, which here are merged to

$$f_{oV} = f_o \left[1 - \left(\frac{2V_{Ed}}{V_{Rd}} - 1 \right)^2 \right] \qquad (6.38)$$

where V_{Rd} is obtained from *clause 6.2.6(2)*. *Clause 6.2.6(2)*

In practice, instead of applying the reduced yield strength, the calculation is performed applying an effective plate thickness.

6.2.11 Web bearing

Clause 6.2.11 concerns the design of webs subjected to localised forces caused by concentrated loads or reactions applied to a beam. This subject is covered in *clause 6.7.5* for unstiffened and longitudinally stiffened webs. *Clause 6.2.11*
Clause 6.7.5

For a transversely stiffened web, the bearing stiffener, if fitted, should be of class 1 or 2 section. It may be conservatively designed on the assumption that it resists the entire bearing force, unaided by the web, the stiffener being checked as a strut (see *clause 6.3.1*) for out-of-plane column buckling and local squashing, with lateral bending effects allowed for if necessary (see *clause 6.3.2*). For plate girders, see *clause 6.7.8*. *Clause 6.3.1*
Clause 6.3.2
Clause 6.7.8

6.3. Buckling resistance of members

Clause 6.3 covers the buckling resistance of members. Guidance is provided for: *Clause 6.3*

- compression members susceptible to flexural, torsional and torsional–flexural buckling *Clause 6.3.1*
- uniform bending members susceptible to lateral torsional buckling *Clause 6.3.2*
- members subjected to a combination of bending and axial compression *Clause 6.3.3*

For member design, no account need be taken for fastener holes at the member ends.

Clauses 6.3.1 to *6.3.3* are applicable to members, not necessarily defined as those with a constant cross-section along the length of the member (Example 6.11 shows how to calculate the buckling resistance of members with stepwise variable cross-section and axial force). For members with tapered sections, Eurocode 9 provides no design expressions for calculating buckling resistances; it is, however, noted that a second-order analysis using the member imperfections according to *clause 5.3.4* may be used to directly determine member buckling resistances. *Clauses 6.3.1–6.3.3*

Clause 5.3.4

6.3.1 Members in compression

General

The Eurocode 9 approach to determining the buckling resistance of compression members is based on the same principles as that of Eurocode 3 for steel. The primary differences between the two codes are that the buckling curves do not depend on the shape of the cross-section (as there are very small residual stresses in aluminium profiles) but on the shape of the stress–strain curve (as it is curved) and that the softening of the material in the HAZ must be allowed for.

Buckling resistance

The design compression force is denoted N_{Ed} (axial design effect). This must be shown to be less than or equal to the design buckling resistance of the compression member, $N_{b,Rd}$ (axial buckling

Clause 6.3.1.1

resistance). See *clause 6.3.1.1*. It is not stated in Eurocode 9 Part 1-1 that members with non-symmetric class 4 cross-sections have to be designed for combined bending and axial compression because of the additional bending moments that result from the shift in neutral axis from the gross cross-section to the effective cross-section (due to local buckling and/or HAZ).

Clause 6.1.3
Clause 6.3.3

However, this is stated for structural sheeting in Part 1-4 *clause 6.1.3*. The design of members subjected to combined bending and axial compression is covered in *clause 6.3.3*.

Clause 6.3.1

Compression members with class 1, 2, 3 and 4 cross-sections follow the provisions of *clause 6.3.1*, where the design buckling resistance should be taken as the lesser of

$$N_{b,Rd} = \kappa \chi \omega_x A_{eff} f_o / \gamma_{M1} \tag{6.49}$$

$$N_{b,Rd} = \chi_{haz} \omega_{x,haz} A_{u,eff} f_u / \gamma_{M2} \qquad \text{in a section with transverse weld} \tag{6.49b}$$

where:

χ is the reduction factor for the relevant buckling mode (flexural, torsional or torsional–flexural). These buckling modes are discussed later in this section.

Clause 6.3.3.3(3)

χ_{haz} is the reduction factor based on $\bar{\lambda}_{haz}$ according to *clause 6.3.3.3(3)*.

κ is a factor to allow for the weakening effects of longitudinal welds. If there are no welds, then $\kappa = 1$, otherwise κ is given by expressions D6.13 and D6.14 from *Table 5.6*, and is dependent on material buckling class BC according to *Table 3.2a* or *3.2b*.

Buckling class A:

$$\kappa = 1 - \left(1 - \frac{A_1}{A}\right)10^{-\bar{\lambda}} - \left(0.05 + 0.1\frac{A_1}{A}\right)\bar{\lambda}^{1.3(1-\bar{\lambda})} \tag{D6.13}$$

where $A_1 = A - A_{haz}(1 - \rho_{o,haz})$ in which A_{haz} = area of the HAZ.

Buckling Class B:

$$\kappa = 1 + 0.04(4\bar{\lambda})^{(0.5-\bar{\lambda})} - 0.22\bar{\lambda}^{1.4(1-\bar{\lambda})} \qquad \text{but } \kappa = 1 \text{ if } \bar{\lambda} \leq 0.2 \tag{D6.14}$$

where:

$\kappa = 1$ for torsional and torsional–flexural buckling and also for members with longitudinal welds.

A_{eff} is the effective area allowing for local buckling and HAZ softening of longitudinal welds. For torsional and torsional-flexural buckling, see *Table 6.7*, referred to later in this guide. For class 1, 2 and 3 cross-sections without longitudinal welds, A_{eff} is the gross cross-section area A_g.

$A_{u,eff}$ is the effective area allowing for local buckling and HAZ softening according to

Clause 6.2.4(2)

        clause 6.2.4(2).

ω_x is the factor allowing for the location of the design section along the member, see

Clause 6.3.3.5

        clause 6.3.3.5. Usually, $\omega_x = 1$ if there are axial force only.

Clause 6.3.3.3
Clause 6.3.3.4

$\omega_{x,haz}$ is the factor allowing for the location of localised weld along the member (see *clause 6.3.3.3*) or localised reduction of the cross-section (see *clause 6.3.3.4*).

Buckling curves

In contrast to steel, the choice of buckling curve is not dependent on the shape of the cross-section. The reason for this is that there are very small residual stresses in extruded aluminium profiles. Instead, the choice of buckling curve is dependent on the shape of the stress–strain curve of the material (Mazzolani, 1995, 2003; Mazzolani *et al.*, 1996). Materials with a large proportional limit are more favourable for buckling than materials with a more rounded-off stress–strain curve. The materials are therefore grouped into two classes (A and B) in Table

Clause 6.3.1.2

3.1 (*Tables 3.2a* and *3.2b*), and two corresponding buckling curves are given in *clause 6.3.1.2*, defined by the imperfection factor α and the limit of the horizontal plateau $\bar{\lambda}_0$.

For flexural buckling:

- buckling class A $\alpha = 0.20$ and $\bar{\lambda}_0 = 0.10$
- buckling class B $\alpha = 0.32$ and $\bar{\lambda}_0 = 0$.

For torsional and torsional–flexural buckling:

- with a general cross-section $\alpha = 0.35$ and $\bar{\lambda}_0 = 0.4$; $A_{\text{eff}} = A_{\text{eff}}$
- composed entirely of radiating outstands $\alpha = 0.20$ and $\bar{\lambda}_0 = 0.6$; $A_{\text{eff}} = A$.

To determine whether a cross-section is 'general' or not the following definitions are given in *Table 6.7*.

- **General**: for sections containing reinforced outstands such that mode 1 (distortional buckling of stiffener) would be critical in terms of local buckling (see *clause 6.1.4.3*), the member should be regarded as 'general', and A_{eff} determined allowing for either or both local buckling and HAZ material.

Clause 6.1.4.3

- **Composed entirely of radiating outstands**: for sections such as angles, tees and cruciforms, local and torsional buckling are closely related. When determining A_{eff}, allowance should be made, where appropriate, for the presence of HAZ material (due to longitudinal welds), but no reduction should be made for local buckling (i.e. $\rho_{\text{c}} = 1$).

The formulation of the buckling curves is according to *clause 6.3.1.2*:

Clause 6.3.1.2

$$\chi = \frac{1}{\phi + \sqrt{\phi^2 - \bar{\lambda}^2}} \qquad \text{but } \chi \leq 1.0 \qquad\qquad (6.50)$$

where

$$\phi = 0.5(1 + \alpha(\bar{\lambda} - \bar{\lambda}_0) + \bar{\lambda}^2)$$

Slenderness
The slenderness $\bar{\lambda}$ (in EN 1999 denoted the slenderness parameter, in EN 1993 the relative slenderness or non-dimensional slenderness ratio, in this guide just slenderness) is defined as

$$\bar{\lambda} = \sqrt{\frac{A_{\text{eff}} f_{\text{o}}}{N_{\text{cr}}}} \qquad \text{but for members with transverse welds, see Section 6.3.3.3 of this guide}$$

$$(6.51)$$

where N_{cr} is the elastic critical force for the relevant buckling mode based on the gross cross-sectional properties. See also Figures 6.29 and 6.30.

Figure 6.29. Reduction factor χ for flexural buckling. (Reproduced from EN 1999-1-1 (*Figure 6.11*), with permission from BSI)

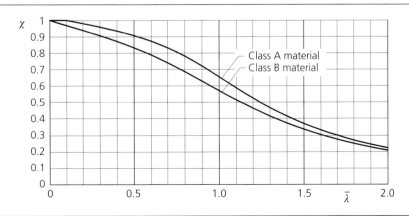

Figure 6.30. Reduction factor χ for torsional and torsional–flexural buckling: 1, cross-section composed of radiating outstands; 2, general cross-section. (Reproduced from EN 1999-1-1 (*Figure 6.12*), with permission from BSI)

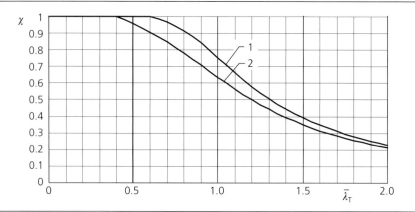

For flexural buckling, the slenderness (*expression 6.51*) can be reformulated as

$$\bar{\lambda} = \sqrt{\frac{A_{\mathrm{eff}} f_{\mathrm{o}}}{N_{\mathrm{cr}}}} = \sqrt{\frac{A_{\mathrm{eff}} f_{\mathrm{o}}}{\pi^2 EI/l_{\mathrm{cr}} 2}} = \frac{l_{\mathrm{cr}}}{i} \frac{1}{\pi} \sqrt{\frac{f_{\mathrm{o}}}{E}} \qquad (D6.15)$$

as

$$N_{\mathrm{cr}} = \frac{\pi^2 EI}{l_{\mathrm{cr}}^2} \quad \text{and} \quad i = \sqrt{\frac{I}{A}} \quad \text{(radius of gyration)} \qquad (D6.16)$$

The slenderness $\bar{\lambda} = \bar{\lambda}_{\mathrm{T}}$ for torsional and torsional–flexural buckling should be taken as

$$\bar{\lambda}_{\mathrm{T}} = \sqrt{\frac{A_{\mathrm{eff}} f_{\mathrm{o}}}{N_{\mathrm{cr}}}} \qquad (6.53)$$

where A_{eff} is the cross-section area according to *Table 7.6*, and N_{cr} is the elastic critical load for torsional buckling, allowing for interaction with flexural buckling if necessary (torsional–flexural buckling). Values of N_{cr} and $\bar{\lambda}_{\mathrm{T}}$ are given in *Annex I*.

Buckling length
There is usually some degree of flexibility in the connections at the ends of a member. Eurocode 9 therefore recommends effective (or buckling) lengths that are larger than the theoretical values for rigid connections. *Table 6.8* provides the buckling length factor k for members with different end conditions illustrated in Figure 6.31, where L is the system length.

Clause 6.3.1.5

For angles, channels and T sections (such as web members in trusses) connected through one leg, web or flange only, a simplified approach is given in ***clause 6.3.1.5***.

Figure 6.31. Recommended buckling length for compression members (*Table 6.8*)

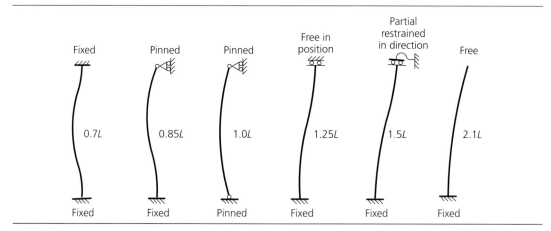

Example 6.10: buckling resistance of a compression member

A circular hollow section member is to be used as a column under a canopy. The column is free at the top and fixed at the base. The column height is $L = 2.4$ m, as shown in Figure 6.32. The vertical loading from gravity and snow load is $N_{Ed} = 50$ kN.

The outer diameter of the section is 120 mm and the thickness is 4 mm, which means the mean radius $r = 60 - 2 = 58$ mm. The material is EN AW-6063 T6/ET which, from Table 3.1 (*Table 3.2b*), has the strengths $f_o = 160$ MPa and $f_u = 195$ MPa. Partial safety factors are $\gamma_{M1} = 1.1$ and $\gamma_{M2} = 1.25$ according to Table 6.1.

Section properties

$$A = 4 \times 2 \times \pi \times 58 = 1458 \text{ mm}^2$$

$$I = \frac{\pi}{64}\left[120^4 - (120 - 2 \times 4)^4\right] = 2.455 \times 10^6 \text{ mm}^4$$

Cross-section classification under axial compression (*clause 6.1.4*)
From Table 6.4 (*Table 6.2*):

Clause 6.1.4

$$\varepsilon = \sqrt{250/f_o} = \sqrt{250/160} = 1.25$$

For a circular hollow section the slenderness β is given by *expression 6.10*:

Clause 6.1.4.3

$$\beta = 3\sqrt{D/t} = 3\sqrt{116/4} = 16.2 \qquad (6.10)$$

Limits for classes 1 and 2 in Table 6.4 (*Table 6.2*):

$$\beta_1 = 11\varepsilon = 13.8 < \beta$$

$$\beta_2 = 16\varepsilon = 20.0 > \beta$$

The section is class 2.

Buckling resistance if the column is fixed into a concrete foundation column (Figure 6.32c, *clause 6.3.1*)
The buckling length is at least 2.1 times the column height according to Figure 6.31 in the last case (*Table 6.8*, case 6):

Clause 6.3.1

$$l_{cr} = 2.1 \times 2400 = 5040 \text{ mm}$$

Figure 6.32. Column with alternative column bases

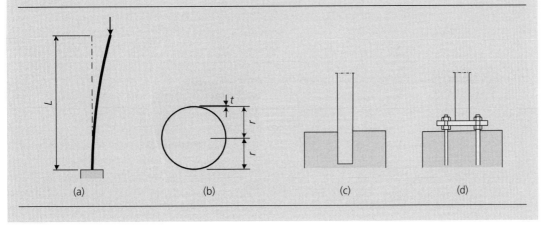

(a)   (b)   (c)   (d)

Clause 6.3.1.2

The buckling load and the slenderness will be, according to *clause 6.3.1.2*,

$$N_{cr} = \frac{\pi^2 EI}{l_{cr}^2} = \frac{\pi^2 \times 70000 \times 2.455 \times 10^6}{5040^2} = 66.77 \, \text{kN}$$

$$\bar{\lambda} = \sqrt{\frac{A_{eff} f_o}{N_{cr}}} = \sqrt{\frac{1458 \times 160}{66770}} = 1.869 \qquad (6.51)$$

The reduction factor for flexural buckling with $\alpha = 0.2$ and $\bar{\lambda}_0 = 0.1$ from *Table 6.6* for buckling class A is

$$\phi = 0.5\left(1 + \alpha(\bar{\lambda} - \bar{\lambda}_0) + \bar{\lambda}^2\right) = 0.5\left(1 + 0.2(1.869 - 0.1) + 1.869^2\right) = 2.424$$

$$\chi = \frac{1}{\phi + \sqrt{\phi^2 - \bar{\lambda}^2}} = \frac{1}{2.424 + \sqrt{2.424^2 - 1.869^2}} = 0.252 \qquad (6.50)$$

The buckling resistance (*expression 6.49*) for $\kappa = 1$, no welds, is

$$N_{b,Rd} = \kappa \chi A_{eff} f_o / \gamma_{M1} = 1.0 \times 0.252 \times 1458 \times 160/1.1 = 53.5 \, \text{kN} > 50 \, \text{kN} \qquad (6.49)$$

which is acceptable.

Clause 6.3.1

Buckling resistance if the column is welded to a plate (Figure 6.32d, *clause 6.3.1*)
Simple conservative method. Check as if the whole column is made of HAZ softened material. From Table 3.1 (*Table 3.2b*), $\rho_{o,haz} = 0.41$, $f_o = \rho_{o,haz} f_o$ and $A_{eff} = A_g$.

$$\bar{\lambda} = \sqrt{\frac{A_{eff} f_o}{N_{cr}}} = \sqrt{\frac{A_g \rho_{o,haz} f_o}{N_{cr}}} = \sqrt{\frac{1458 \times 0.41 \times 160}{66770}} = 1.197 \qquad (6.51)$$

The reduction factor for flexural buckling, from *expression 6.50* (with $\alpha = 0.2$ and $\bar{\lambda}_0 = 0.1$ from *Table 6.6* for buckling class A), is $\chi = 0.527$.

The buckling resistance (*expression 6.49*) for $\kappa = 1$ (no longitudinal welds) is

$$N_{b,Rd} = \kappa \chi f_o A_{eff} / \gamma_{M1} = 1 \times 0.527 \times 0.41 \times 160 \times 1458/1.1 = 45.8 \, \text{kN} < 50 \, \text{kN} \qquad (6.49)$$

which is not acceptable.

Buckling resistance if using expression 6.49b for the section with localised weld (*clause 6.3.1.1*)

Clause 6.3.1.1
Clause 6.3.1.1(2)

According to *clause 6.3.1.1(2)*, in a section with transverse weld

$$N_{b,Rd} = \chi_{haz} \omega_{x,haz} A_{u,eff} f_u / \gamma_{M2} \qquad (6.49b)$$

Clause 6.3.3.3

where $\omega_{x,haz}$ is given in *clause 6.3.3.3*, which also states that the **ultimate strength** should be used in the HAZ.

From Table 3.1 (*Table 3.2b*), $\rho_{u,haz} = 0.56$. As the column is welded all around the periphery to the column plate, the whole section is HAZ softened, so

$$A_{u,eff} = \rho_{u,haz} A = 0.56 \times 1458 = 816 \, \text{mm}^2$$

Clause 6.3.3.3(3)

The reduction factor χ_{haz} is based on the slenderness $\bar{\lambda}_{haz}$ according to *clause 6.3.3.3(3)*.

$$\bar{\lambda}_{haz} = \sqrt{\frac{A_{u,eff} f_u \gamma_{M1}}{N_{cr} \gamma_{M2}}} = \sqrt{\frac{816 \times 195 \times 1.1}{66770 \times 1.25}} = 1.448 \qquad (6.67)$$

The reduction factor for flexural buckling with $\alpha = 0.2$ and $\bar{\lambda}_0 = 0.1$ from *Table 6.6* for buckling class A is $\chi_{haz} = 0.393$ (*expression 6.50*).

The distance from the top to the weld section is the column height 2400 mm, and the buckling length is 5040 mm. The factor $\omega_{x,haz}$ as given by *expression 6.65* in **clause 6.3.3.3(2)** is then

Clause 6.3.3.3(2)

$$\omega_{x,haz} = \frac{1}{\chi_{haz} + (1 - \chi_{haz})\sin(\pi x_{s,haz}/l_{cr})} = \frac{1}{0.393 + (1 - 0.393)\sin(\pi \times 2400/5040)}$$
$$= 1.002 \tag{6.65}$$

In this case, the section with the weld is very close to the middle of the buckling length, so the sine value is close to 1, and therefore $\omega_{x,haz}$ is also close to 1.

The buckling resistance (*expression 6.49b*) is

Clause 6.3.1.1

$$N_{b,Rd} = \chi_{haz}\omega_{x,haz}A_{u,haz}f_u/\gamma_{M2} = 0.393 \times 1.002 \times 816 \times 195/1.25$$
$$= 50.1\,\text{kN} > 50\,\text{kN} \tag{6.49}$$

which is acceptable.

Example 6.11: buckling resistance of a member with a stepwise variable cross-section

To illustrate the use of the code for arbitrary compression members, a cantilever with a stepwise variable rectangular hollow section is checked for stepwise variable axial force (Figure 6.33). In an actual structure the column should be loaded with a bending moment as well, but this is omitted in this example. Beam columns are treated later in this guide.

Figure 6.33. Cantilever column with a stepwise variable cross-section

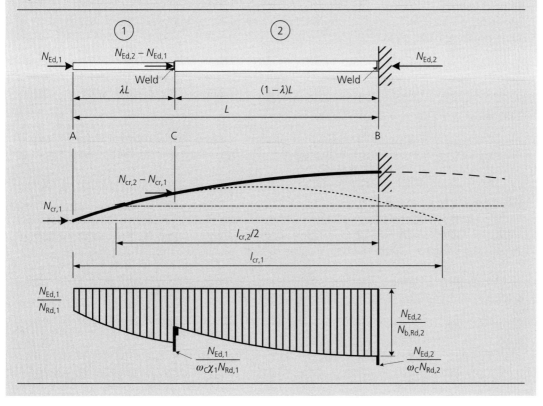

The cantilever is free at A and fixed at B, and has a step in the cross-section at C that is $\lambda L = 0.3L$ from A. The span is $L = 5.0$ m. The load at A is $N_{Ed,1} = 100$ kN, and at B it is $N_{Ed,2} = 250$ kN. The material is EN AW-6005A T6 with proof strength $f_o = 215$ MPa from Table 3.1 (*Table 3.2b*).

Cross-section

The cross-section area of the two parts are $A_1 = 3880$ mm^2 and $A_2 = 9100$ mm^2, and the second moments of the area are $I_1 = 16.73 \times 10^6$ mm^4 and $I_2 = 80.91 \times 10^6$ mm^4. The cross-section is class 2.

Buckling lengths

The buckling lengths may be found in handbooks, using the finite-element method or, as in this case, from the equation (Höglund, 1968)

$$\tan(k_2(1 - \lambda)L)\tan\left(\sqrt{\frac{N_1 I_2}{N_2 I_1}} k_2 \lambda L\right) = \sqrt{\frac{N_2 I_2}{N_1 I_1}} \tag{D6.17}$$

or, with inserted values,

$$\tan(k_2(1 - 0.3)L)\tan(1.391 \times k_2 \times 0.3L) = 3.477 \tag{D6.18}$$

The solution is

$$k_2 L = 1.853 \tag{D6.19}$$

As

$$k_2^2 = \frac{N_{cr,2}}{EI_2} = \frac{\pi^2 EI_2}{l_{cr,2}^2 EI_2} = \frac{\pi^2}{l_{cr,2}^2} \tag{D6.20}$$

the buckling length for part 2 is

$$l_{cr,2} = \frac{\pi}{k_2} = \frac{\pi L}{1.853} = \frac{\pi \times 5000}{1.853} = 8477 \text{ mm} \tag{D6.21}$$

As the ratio between the buckling loads for the two parts is the same as the ratio between the loads on these parts, then

$$\frac{N_2}{N_1} = \frac{N_{cr,2}}{N_{cr,1}} = \frac{\pi^2 EI_2}{l_{cr,2}^2} \frac{l_{cr,1}^2}{\pi^2 EI_1}$$

from which for part 1

$$l_{cr,1} = \sqrt{\frac{I_1 N_2}{I_2 N_1}} l_{cr,2} = \sqrt{\frac{16.73 \times 250}{80.91 \times 100}} = 8477 = 6095 \text{ mm} \tag{D6.22}$$

Clause 6.3.1.2

Slenderness (*clause 6.3.1.2*)

The slenderness for the two parts at the welded sections C and B are dependent on the reduction factor for the HAZ according to *expression 6.64*. As the section is welded all around the periphery, then the whole cross-section is heat affected, and $A_{u,eff} = \rho_{u,haz} A$.

Clause 6.3.3.3(3)
Clause 6.3.1.2(1)

The resulting slenderness and reduction factors according to *expression 6.50* in **clause 6.3.3.3(3)** and *expression 6.67* in **clause 6.3.1.2(1)** are

$$\bar{\lambda}_{1,haz} = \sqrt{\frac{\rho_{u,haz} A_1 f_u l_{cr,1}^2}{\pi^2 EI_1} \frac{\gamma_{M1}}{\gamma_{M2}}} = \sqrt{\frac{0.63 \times 3880 \times 260 \times 6095^2 \times 1.1}{\pi^2 \times 70\,000 \times 16.73 \times 10^6 \times 1.25}} = 1.341 \rightarrow \chi_1 = 0.445$$

$$(5.50), (6.67)$$

$$\bar{\lambda}_{2,\text{haz}} = \sqrt{\frac{\rho_{\text{u,haz}} A_2 f_\text{u} l_{\text{cr},2}^2}{\pi^2 EI_2} \frac{\gamma_\text{M1}}{\gamma_\text{M2}}} = \sqrt{\frac{0.63 \times 9100 \times 260 \times 8477^2 \times 1.1}{\pi^2 \times 70\,000 \times 80.91 \times 10^6 \times 1.25}} = 1.181 \rightarrow \chi_2 = 0.537$$

$$(6.50),\ (6.67)$$

Part 1, section C. Section C is located at $x_{\text{s,haz}} = \lambda L = 0.3L$ from the end of the equivalent column for part 1. The factor $\omega_{\text{x,haz}}$ for the HAZ in section C is, according to *expression 6.65)* in **clause 6.3.3.3**,

Clause 6.3.3.3

$$\omega_{\text{C,haz}} = \frac{1}{\chi_{1,\text{haz}} + (1 - \chi_{1,\text{haz}}) \sin(\pi\lambda L/l_{\text{cr},1})}$$

$$= \frac{1}{0.445 + (1 - 0.455) \sin(\pi \times 1500/6095)} = 1.201 \qquad (6.65)$$

The buckling resistance for part 1 is, according to *expression 6.49b* in **clause 6.3.1.1**,

Clause 6.3.1.1

$$N_{\text{b,Rd,1}} = \chi_{1,\text{haz}} \omega_{\text{C,haz}} \rho_{\text{u,haz}} A_1 f_\text{u}/\gamma_\text{M2}$$

$$= 0.445 \times 1.201 \times 0.63 \times 3880 \times 260/1.2 = 272\,\text{kN} \qquad (6.49b)$$

The utilisation grade for the axial force is denoted K (B for the bending moment, see Section 6.3.3), and in the HAZ at C it is

$$K_\text{C} = \frac{N_{\text{Ed,1}}}{N_{\text{b,Rd,1}}} = \frac{100}{272} = 0.368 \qquad (D6.23)$$

Part 2, Section B. Section B is in the middle of the equivalent column with buckling length $l_{\text{cr},2} = 8477\,\text{mm}$. The sine term in *expression 6.65* in **clause 6.3.3.3** is 1.0, so $\omega_{\text{B,haz}} = 1$.

Clause 6.3.3.3

The buckling resistance for part 2 is, according to *expression 6.49*,

$$N_{\text{b,Rd,2}} = \chi_{2,\text{haz}} \omega_{\text{B,haz}} \rho_{\text{u,haz}} A_2 f_\text{u}/\gamma_\text{M2}$$

$$= 0.537 \times 1 \times 0.63 \times 9100 \times 260/1.25 = 641\,\text{kN} \qquad (6.49)$$

The utilisation grade in the HAZ at B is

$$K_\text{B} = \frac{N_{\text{Ed,2}}}{N_{\text{b,Rd,2}}} = \frac{250}{641} = 0.390 \qquad (D6.24)$$

The utilisation grade along the column is sketched in Figure 6.33, where K_C and K_B in the HAZs are marked with short lines. The cross-section resistances in sections without HAZs, ignoring second-order bending moments, are

$$N_{\text{Rd,1}} = A_1 f_\text{o}/\gamma_\text{M1} = 3880 \times 215/1.1 = 758\,\text{kN}$$

so $K_\text{A} = N_{\text{Ed,1}}/N_{\text{Rd,1}} = 100/758 = 0.132$ in part 1 at A in unaffected material, and

$$N_{\text{Rd,2}} = A_2 f_\text{o}/\gamma_\text{M1} = 9100 \times 215/1.1 = 1779\,\text{kN}$$

so $K_{\text{C,2}} = N_{\text{Ed,2}}/N_2 = 250/1779 = 0.140$ in part 2 at C in unaffected material.

Design of welds at the splice and the joint
The welded splice at C and the joint at B should be designed not only for the axial force but also for a second-order bending moment according to expression D6.33 (see 'Derivation of formula for beam-column with end moments and/or transverse loads' in Section 6.3.3.1 of this guide):

$$\Delta M = \frac{NW}{A} \left(\frac{1}{\chi} - 1\right) \sin\left(\frac{\pi x}{l_{\text{cr}}}\right)$$

At the splice C, $W_1 = 2.57 \times 10^5 \, \text{mm}^3$, and

$$\Delta M_C = \frac{N_{Ed,1} W_1}{A_1} \left(\frac{1}{\chi_{1,haz}} - 1 \right) \sin\left(\frac{\pi \lambda L}{l_{cr,1}} \right)$$

$$= \frac{100 \times 2.57 \times 10^5}{3880} \left(\frac{1}{0.445} - 1 \right) \sin\left(\frac{\pi \times 1500}{6095} \right)$$

$$= 5767 \, \text{kN mm} = 5.77 \, \text{kN m} \tag{D6.25a}$$

At the fixed end B, $W_2 = 9.00 \times 10^5 \, \text{mm}^3$, and

$$\Delta M_B = \frac{N_{Ed,2} W_2}{A_2} \left(\frac{1}{\chi_{2,haz}} - 1 \right) \sin\left(\frac{\pi}{2} \right)$$

$$= \frac{250 \times 9.00 \times 10^5}{9100} \left(\frac{1}{0.537} - 1 \right) 1.0 = 2.13 \times 10^7 \, \text{kN mm} = 21.3 \, \text{kN m} \tag{D6.25b}$$

6.3.2 Members in bending

General

Laterally unrestrained beams subject to bending about their major axis have to be checked for lateral torsional buckling in accordance with *clause 6.3.2*. As described in Section 6.2.5 of this guide, there are a number of common situations where lateral torsional buckling need not be considered, and member strength may be assessed on the basis of the in-plane cross-sectional resistance.

Clause 6.3.2

In *Annex I* there are design aids that simplify calculations. Formulae and tables to calculate the elastic critical moment are given, but also formulae and tables for the direct calculation of the slenderness $\bar{\lambda}_{LT}$ (*clause I.4*).

Clause I.4

Effective lateral restraint

Clause 6.3.2.1

A note in *clause 6.3.2.1* deems that '*lateral torsional buckling need not be checked ... if the member is fully restrained against lateral movement through its length*'. Bracing systems providing lateral restraint should be designed according to *clause 5.3.3*.

Clause 5.3.3

Where a series of two or more parallel members require lateral restraint, restraint should be provided by anchoring the ties to an independent robust support, or by providing a triangulated bracing system. If there are many parallel members, it is sufficient for the restraint system to be designed to resist a reduced sum of the lateral forces according to *clause 5.3.3*.

Clause 5.3.3

Further guidance on lateral restraint is available (Nethercot and Lawson, 1992).

Lateral torsional buckling resistance

The design bending moment is denoted M_{Ed} (the bending moment design effect) and the lateral torsional buckling resistance by $M_{b,Rd}$ (the buckling resistance moment). Clearly, M_{Ed} must be shown to be less than $M_{b,Rd}$, and checks should be carried out on all unrestrained segments of beams between points where lateral restraints exists.

The design buckling resistance of a laterally unrestrained beam, or segment of a beam, should be taken as the lesser of

$$M_{b,Rd} = \chi_{LT} \omega_{xLT} \alpha W_{el,y} f_o / \gamma_{M1} \tag{6.55}$$

$$M_{b,Rd} = \chi_{LT,haz} \omega_{xLT,haz} W_{u,eff} f_u / \gamma_{M2} \quad \text{at sections with localised transverse weld} \tag{6.55b}$$

where

$W_{el,y}$ is the elastic section modulus of the gross section, without reduction for HAZ softening, local buckling or holes.

α	is the shape factor taken from Table 6.8 (*Table 6.4* in *clause 6.2.5.1*) subject to the limitation $\alpha \leq W_{pl,y}/W_{el,y}$.	*Clause 6.2.5.1*
$W_{u,eff}$	is the section modulus allowing for local buckling and HAZ softening according to *clause 6.2.5.1(2)*.	*Clause 6.2.5.1(2)*
χ_{LT}	is the reduction factor for lateral torsional buckling (see *clause 6.3.2.2*).	*Clause 6.3.2.2*
$\chi_{LT,haz}$	is the reduction factor for lateral torsional buckling based on $\overline{\lambda}_{LT,haz}$ according to *clause 6.3.3.3(3)*.	*Clause 6.3.3.3(3)*
ω_{xLT}	is the factor allowing for the location of the design section along the member (see *clause 6.3.3.5*). Conservatively, $\omega_{xLT} = 1$.	*Clause 6.3.3.5*
$\omega_{xLT,haz}$	is the factor allowing for the location of transverse weld along the member (see *clause 6.3.3.3*) or a localised reduction of the cross-section (see *clause 6.3.3.4*).	*Clause 6.3.3.3* *Clause 6.3.3.4*

Lateral torsional buckling curves

The reduction factor for lateral torsional buckling χ_{LT} for the appropriate slenderness $\overline{\lambda}_{LT}$ should be determined from *expression 6.56* in *clause 6.3.2.2*:

Clause 6.3.2.2

$$\chi_{LT} = \frac{1}{\phi_{LT} + \sqrt{\phi_{LT}^2 - \overline{\lambda}_{LT}^2}} \quad \text{but } \chi_{LT} \leq 1.0 \tag{6.56}$$

where

$$\phi_{LT} = 0.5\left[1 + \alpha_{LT}(\overline{\lambda}_{LT} - \overline{\lambda}_{0,LT}) + \overline{\lambda}_{LT}^2\right] \tag{6.57}$$

The imperfection factor α_{LT} and the limit of the horizontal plateau $\overline{\lambda}_{0,LT}$ should be taken as $\alpha_{LT} = 0.10$ and $\overline{\lambda}_{0,LT} = 0.6$ for class 1 and 2 cross-sections, and $\alpha_{LT} = 0.20$ and $\overline{\lambda}_{0,LT} = 0.4$ for class 3 and 4 cross-sections. $\overline{\lambda}_{LT}$ is the slenderness for lateral torsional buckling, see later in this guide.

Values of the reduction factor χ_{LT} for the appropriate slenderness $\overline{\lambda}_{LT}$ may be obtained from Figure 6.34 (*Figure 6.13*).

The slenderness $\overline{\lambda}_{LT}$ should be determined from

$$\overline{\lambda}_{LT} = \sqrt{\frac{\alpha W_{el,y}}{M_{cr}}} \quad \text{but for members with transverse welds, see } clause\ 6.3.3.3(3) \tag{6.58}$$

Clause 6.3.3.3(3)

where α is the shape factor given after *expression 6.55*, and M_{cr} is the elastic critical moment for lateral torsional buckling. M_{cr} is based on gross cross-sectional properties, and takes into account the loading conditions, the moment distribution and the lateral restraints. Expressions for M_{cr} for certain sections and boundary conditions are given in *Annex I, clause I.1*, and approximate values of $\overline{\lambda}_{LT}$ for certain I sections and channels are given in *clause I.2*.

Clause I.1
 Clause I.2

Figure 6.34. Reduction factor for lateral-torsional buckling. (Reproduced from EN 1999-1-1 (*Figure 6.13*), with permission from BSI)

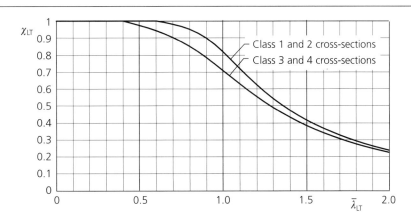

Example 6.12: lateral torsional buckling resistance

A simply supported primary beam supports three secondary beams, as shown in Figure 6.35. Full lateral restraint is assumed at the load application points B, C and D. Check the beam with flange lips for the loads $F_{Ed} = 35\,kN$.

The loading, shear force and bending moment diagrams are shown in Figure 6.35. A lateral torsional buckling check will be carried out on segment BC. Shear and patch loading checks are omitted in this example.

Segment length, cross-section properties, material and bending moments

Segment length	$l_{segm} = 2000\,mm$
Section height	$h_e = 300\,mm$
Flange width	$b_e = 160\,mm$
Web thickness	$t_w = 10\,mm$
Flange thickness	$t_f = 10\,mm$
Flange lip	$c_e = 35\,mm$
Web height	$b_w = h_e - 2t_f = 280\,mm$
EN-AW 6082-T6	$f_o = 260\,MPa$
Partial safety factor	$\gamma_{M1} = 1.1$

Bending moment at end C:

$$M_{C,Ed} = 1.5F_{Ed}2l_{segm} - F_{Ed}l_{segm} = 2 \times 35 \times 2 = 140\,kN\,m$$

Bending moment at B and D:

$$M_{B,Ed} = 1.5F_{Ed}l_{segm} = 1.5 \times 35 \times 2 = 105\,kN\,m$$

Clause 6.1.4

Cross-section classification in y–y axis bending (*clause 6.1.4*)

$$\varepsilon = \sqrt{250/f_o} = \sqrt{250/260} = 0.981$$

Limits for classes 1 and 2 for outstands:

$$\beta_{1,o} = 3\varepsilon = 2.94$$

$$\beta_{2,o} = 4.5\varepsilon = 4.41$$

$$\beta_{3,o} = 6\varepsilon = 5.88$$

Figure 6.35. General arrangement – loading and cross-section

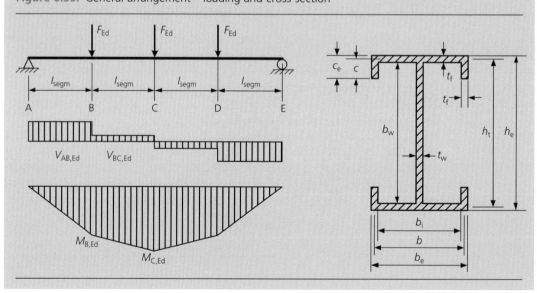

Limits for classes 1 and 2 for internal parts:

$$\beta_{1,i} = 11\varepsilon = 10.8$$

$$\beta_{2,i} = 16\varepsilon = 15.7$$

Outstand flange lip (*expression 6.1*):

$$\beta_c = (c_e - t_f)/t_f = (35 - 10)/10 = 2.50 \rightarrow \text{class 1}$$

Internal flange parts (*expression 6.1*):

$$\beta_f = (0.5b - 0.5t_w - t_f)/t_f = (80 - 5 - 10)/10 = 6.50 \rightarrow \text{class 1}$$

Flange with lip (*expressions 6.6 and 6.7a*):

$$\beta = \eta\frac{b}{t} = \frac{1}{\sqrt{1 + 0.1(c/t - 1)^2}}\frac{b}{t} = \frac{1}{\sqrt{1 + 0.1[(35-10)/10-]^2}}\frac{80 - 5 - 10}{10} = 5.87 \rightarrow \text{class 3}$$

Web – internal part (*expression 6.1*):

$$\beta_w = 0.4b_w/t_w = 0.4 \times 272/10 = 10.9 \rightarrow \text{class 2}$$

In y–y axis bending, the overall cross-section classification is class 3. As $\beta = 5.87$ for the flange with a lip is very close to the limit $\beta_{3,o} = 5.88$ for class 4, the resistance is based on the elastic section modulus, and the shape factor is $\alpha = 1.0$.

Design resistance for *y*–*y* axis bending (*clause 6.2.5*)
The second moment of the area is

$$I_y = \frac{1}{12}\left(160 \times 300^3 - 130 \times 280^3 - 20 \times 230^3\right) = 1.019 \times 10^8 \text{ mm}^4$$

$$W_{el,y} = \frac{1.019 \times 10^8}{150} = 6.794 \times 10^5 \text{ mm}^3$$

The cross-section resistance for y–y axis bending is

$$M_{y,Rd} = \alpha W_{el,y}f_o/\gamma_{M1} = 1 \times 6.794 \times 260 \times 10^5/1.1 = 160.6 \text{ kN m}$$

Lateral torsional buckling of the segment in bending (*clause 6.3.2.1*)
The elastic lateral torsional buckling load is found in *Annex I*. The warping constant and the torsion constant are found in *Annex J*, the torsion constant in ***clause J.1*** and the warping constant in *Figure J.2* (case 8) in ***clause J.3***.

The second moment of area with respect to the z–z-axis:

$$I_z = \frac{1}{12}\left(300 \times 160^3 - (300 - 2 \times 35)160^3 - 2 \times 25 \times 140^3\right) = 1.246 \times 10^7 \text{ mm}^4$$

The warping constant:

$$I_w = \frac{h_f^2 I_z}{4} + \frac{c^2 b^2 t}{6}(3h_f + 2c) = \frac{290^2 \times 1.246 \times 10^7}{4} + \frac{30^2 \times 150^2 \times 10}{6}(3 \times 290 + 2 \times 30)$$

$$= 2.934 \times 10^{11} \text{ mm}^6$$

The torsion constant:

$$I_t = \sum bt^3/3 = \tfrac{1}{3}\left(2 \times 150 \times 10^3 + 280 \times 10^3 + 4 \times 30 \times 10^3\right) = 2.333 \times 10^5 \text{ mm}^4 \quad (J1.a)$$

Clause 6.2.5

Clause 6.3.2.1

Clause J.1
Clause J.3

The elastic critical moment for lateral torsional buckling is given by the general formula

$$M_{cr} = \mu_{cr} \frac{\pi\sqrt{EI_z GI_t}}{L} \qquad (I.2)$$

where the relative non-dimensional critical moment μ_{cr} is

$$\mu_{cr} = \frac{C_1}{k_z} \left(\sqrt{1 + \kappa_{wt}^2 + (C_2 \zeta_g - C_3 \zeta_j)^2} - (C_2 \zeta_g - C_3 \zeta_j) \right) \qquad (I.3)$$

The standard conditions of restraint at each end are used, which means $k_z = 1$, $k_w = 1$ and $k_y = 1$. The non-dimensional torsion parameter is then

$$\kappa_{wt} = \frac{\pi}{k_w L} \sqrt{\frac{EI_w}{GI_t}} = \frac{\pi}{1.0 \times 2000} \sqrt{\frac{70\,000 \times 2.934 \times 10^{11}}{27\,000 \times 2.333 \times 10^5}} = 2.836$$

Clause I.1.2

As $k_z = 1$, the value of C_1 for any ratio of end moment loading, as indicated in *Table I.1* in *clause I.1.2*, is given approximately by *expression I.6*, where $\psi = M_{B,Ed}/M_{C,Ed} = 0.75$:

$$C_1 = (0.310 + 0.428\psi + 0.262\psi^2)^{-0.5} = (0.310 + 0.428 \times 0.75 + 0.262 \times 0.75^2)^{-0.5}$$

$$= 1.133 \qquad (I.6)$$

Values of C_2 and C_3 given in *Tables I.1* and *I.2* are not needed in this case, as $\zeta_g = 0$ and $\zeta_j = 0$.

The relative non-dimensional coordinate of the point of load application is related to the shear centre $\zeta_g = 0$ as well as to the relative non-dimensional cross-section mono-symmetry parameter $\zeta_j = 0$. *Expression I.3* for μ_{cr} is then simplified to

$$\mu_{cr} = \frac{C_1}{k_z} \left[\sqrt{1 + \kappa_{wt}^2} \right] = \frac{1.133}{1} \sqrt{1 + 2.836^2} = 3.408 \qquad (I.3)$$

Now, the elastic critical moment for lateral torsional buckling can be calculated as

$$M_{cr} = \mu_{cr} \frac{\pi\sqrt{EI_z GI_t}}{L} = 3.673 \frac{\pi\sqrt{70\,000 \times 1.246 \times 10^7 \times 27\,000 \times 2.333 \times 10^5}}{2000}$$

$$= 397\,\text{kN m} \qquad (I.2)$$

The slenderness $\bar{\lambda}_{LT}$ is determined from

$$\bar{\lambda}_{LT} = \sqrt{\frac{\alpha W_{el,y} f_o}{M_{cr}}} = \sqrt{\frac{1 \times 6.794 \times 10^5 \times 260}{397 \times 10^6}} = 0.667 \qquad (6.58)$$

Clause 6.3.2.2(1)

For class 1 and 2 cross-sections, the parameters in the formulae for the reduction factor χ_{LT} for lateral torsional buckling are $\alpha_{LT} = 0.20$ and $\bar{\lambda}_{0,LT} = 0.4$ according to *clause 6.3.2.2(1)*:

$$\phi_{LT} = 0.5(1 + \alpha_{LT}(\bar{\lambda}_{LT} - \bar{\lambda}_{0,LT}) + \bar{\lambda}_{LT}^2) = 0.5(1 + 0.2(0.667 - 0.4) + 0.667^2) = 0.749$$

$$(6.57)$$

$$\chi_{LT} = \frac{1}{\phi_{LT} + \sqrt{\phi_{LT}^2 - \bar{\lambda}_{LT}^2}}$$

$$= \frac{1}{0.749 + \sqrt{0.749^2 - 0.667^2}} = 0.917 \quad \text{but } \chi_{LT} \leq 1 \qquad (6.56)$$

The lateral torsional buckling resistance in segment BC is

$$M_{b,Rd} = \frac{\chi_{LT}\alpha W_{el,y}f_o}{\gamma_{M1}} = \frac{0.917 \times 1 \times 6.794 \times 10^5 \times 260}{1.1} = 147\,kN\,m$$

$$\frac{M_{C,Ed}}{M_{b,Rd}} = \frac{145}{147} = 0.985 < 1.0$$

Segment BC is acceptable.

Simplified assessment of slenderness

In *Annex I*, *clause I.2*, a simplified approximate method is provided for calculation of the slenderness $\bar{\lambda}_{LT}$ without calculating the lateral torsional critical moment M_{cr}. For I sections and channels covered by *Table I.5*, the value of $\bar{\lambda}_{LT}$ may be obtained from *expression I.11* with λ_{LT} from *expression I.12* in which X and Y are coefficients obtained from *Table I.5*. For a lipped I-section, case 2, we get (note the notations for h, b and c) *Clause I.2*

$$X = 0.94 - (0.03 - 0.07c/b)h/b - 0.3c/b$$

$$= 0.94 - (0.03 - 0.07 \times 25/160)300/160 - 0.3 \times 25/160 = 0.86$$

$$Y = 0.05 - 0.06c/h = 0.05 - 0.06 \times 25/300 = 0.045$$

The formulae are valid for $1.5 \le h/b \le 4.5$ and $0 \le c/b \le 0.5$, which is fulfilled in this example ($h/b = 300/160 = 1.88$ and $c/b = 25/160 = 0.156$). Note that they are valid for non-lipped sections as well.

$$i_z = \sqrt{\frac{I_z}{A}} = \sqrt{\frac{1.246 \times 10^7}{7000}} = 42.2\,mm$$

$$\lambda_{LT} = \frac{XL/i_z}{\left[1 + Y\left(\dfrac{L/i_z}{h/t_2}\right)\right]^{1/4}} = \frac{0.86 \times 2000/42.2}{\left[1 + 0.045\left(\dfrac{2000/42.2}{300/10}\right)\right]^{1/4}} = 39.6 \qquad (I.12)$$

$$\bar{\lambda}_{LT} = \lambda_{LT}\frac{1}{\pi}\sqrt{\frac{\alpha f_o}{E}} = 39.6\frac{1}{\pi}\sqrt{\frac{1 \times 260}{70\,000}} = 0.768 \qquad (I.11)$$

The reduction factor will be $\chi_{LT} = 0.869$, and the moment resistance $M_{b,Rd} = 139.5\,kN\,m$.

$$\frac{M_{C,Ed}}{M_{b,Rd}} = \frac{140}{139.5} = 1.003 \approx 1.0$$

Segment BC is still acceptable, but the utilisation grade is larger. The obvious reason for this is the fact that the moment gradient is not taken account of in the simplified method.

6.3.3 Members in bending and axial compression

Members subject to bending and axial compression (beam columns) exhibit complex structural behaviour. First-order bending moments about the major and/or minor axes ($M_{y,Ed}$ and $M_{z,Ed}$, respectively) are induced by lateral loading and/or end moments. The addition of the axial loading N_{Ed} not only results in the axial force in the member but also amplifies the bending moments about both principal axes (second-order bending moments). Since, in general, the bending moment distributions about both principal axes will be non-uniform, and hence the most heavily loaded cross-section can occur at any point along the length of the member, the design treatment is usually complex.

Members subject to bending and axial compression are usually parts of a frame structure. Second-order sway effects (P–Δ effects) should be allowed for, either by using suitably enhanced end moments or by using appropriate buckling lengths. The design formulations in

Figure 6.36. Flexural buckling and lateral torsional buckling

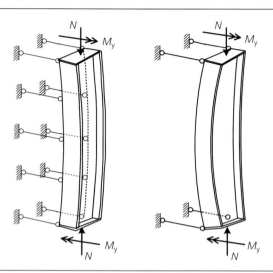

Clause 6.3.3 *clause 6.3.3* imply, in principle, that all sections along the member should be checked, so a cross-section check at each end of the member will be included.

Clause 6.1.4 It should be noted that the classification of cross-sections for members in combined bending and axial compression is made for the loading components separately according to *clause 6.1.4*. No classification is needed for the combined state of stress. This means that a cross-section can belong to different classes for axial force, major axis bending and minor axis bending. The combined state of stress is allowed for in the interaction expressions, which should be used for all classes of cross-section. The influence of local buckling and yielding on the resistance for combined loading is included in the resistances $N_{b,Rd}$, $M_{y,Rd}$ (or $M_{y,b,Rd}$) and $M_{z,Rd}$, and the exponents ξ_{yc}, η_c and ξ_{zc}, which all are functions of the slenderness of the cross-section *Clauses 6.3.3.1–6.3.3.5* and the member. Furthermore, it should be noted that a section check is included in the check of flexural and lateral-torsional buckling if the methods in *clauses 6.3.3.1* to *6.3.3.5* are used.

Two buckling modes are recognised for a beam column in mono-axis (*y*-axis) bending (Figure 6.36):

- flexural buckling – members are not susceptible to torsional deformations or braced in the lateral direction
- lateral torsional buckling – members are susceptible to torsional deformation.

The former is for cases where no lateral torsional buckling is possible, for example for members with square or circular hollow sections, as well as for arrangements where torsional and/or lateral deformation is prevented (see Figure 6.36). Beam columns in minor axis bending and rectangular hollow sections with height/width less than 2 also belong to this category. Most I section columns in building frames are likely to fall within the second category.

6.3.3.1 Flexural buckling
Clause 6.3.3.1 Three formulae are given for members with different cross-sections in *clause 6.3.3.1*:

- Open cross-sections (typically I sections) for major axis (*y* axis) bending

$$\left(\frac{N_{Ed}}{N_{y,b,Rd}}\right)^{\xi_{yc}} + \frac{M_{y,Ed}}{M_{y,Rd}} \leq 1.00 \qquad (6.59)$$

- Symmetric cross-sections (also typically I sections) for minor axis (*z*-axis) bending as well as solid cross-sections

$$\left(\frac{N_{Ed}}{N_{z,b,Rd}}\right)^{\eta_c} + \left(\frac{M_{z,Ed}}{M_{z,Rd}}\right)^{\xi_{zc}} \leq 1.00 \qquad (6.60)$$

■ Hollow cross-sections and tubes

$$\left(\frac{N_{Ed}}{N_{b,Rd,min}}\right)^{\psi_c} + \left[\left(\frac{M_{y,Ed}}{M_{y,Rd}}\right)^{1.7} + \left(\frac{M_{z,Ed}}{M_{z,Rd}}\right)^{1.7}\right]^{0.6} \leq 1.00 \tag{6.62}$$

The exponents are functions of the shape factors and the reduction factor for flexural buckling, or 0.8 for simplicity. See Example 6.14.

Expression 6.59 may also be used for other open single-symmetrical cross-sections, bending about either axis, with appropriate exponents and resistances.

The notations in *expressions 6.59* to *6.62* are as follows:

N_{Ed}	is the design value of the axial compressive force.
$M_{y,Ed}$, $M_{z,Ed}$	are the design values of bending moment about the y–y and z–z axis. The moments are calculated according to **first order theory**.
$N_{y,b,Rd}$, $N_{z,b,Rd}$	are the axial force resistances with respect to the y–y and z–z axis according to *clause 6.3.1* and $N_{b,Rd,min} = \min(N_{y,b,Rd}, N_{z,b,Rd})$.
$M_{y,Rd}$, $M_{z,Rd}$	are the bending moment resistance with respect to the y–y and z–z axis according to *clause 6.2.5*.
α_y, α_z	are the shape factors, but α_y and α_z should not be greater than 1.25. See *clause 6.2.5* and *clause 6.2.9.1(1)*.

Clause 6.3.1

Clause 6.2.5

Clause 6.2.5
Clause 6.2.9.1(1)

It should be noted that all resistances relate to the individual member checks under either compression or bending described in the two previous sections of this guide. The ω factors in the resistances allow for the bending moment distribution along the member or localised transverse HAZs, if any.

Derivation of formula for a beam column with end moments and/or transverse loads

To understand how to use the formulations for the design of beam columns with an arbitrary distribution of bending moments along the beam column, the derivation of the interaction formulae are shown (Höglund, 1968). The derivation is based on elastic theory, where the stress can be given by the well-known expression

$$\sigma(x) = \frac{N}{A} + \frac{M(x)}{W} + \frac{N \times y(x)}{W} \tag{D6.26}$$

where the deflection $y(x)$ is due to the sum of the first-order bending moment $M(x)$ and the additional bending moment $N \times y(x)$ (Figure 6.37). One essential assumption is that the deflection at failure of the beam column is the same as the deflection for the buckling load $N_b = \chi N_o$ only, where $N_o = Af_o$ and χ is the reduction factor for flexural buckling.

Figure 6.37. First- and second-order bending moments

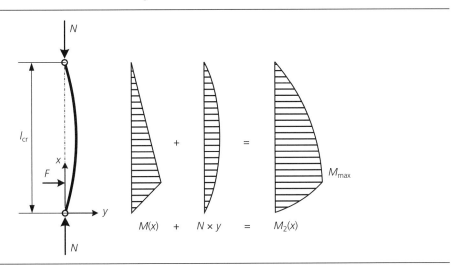

This assumption may seem to be rather rough, but the result has been found to be in accordance with more accurate theories and tests. The reason is that when the axial force N is large, then $M(x)$ is small so the assumed deflection curve is very accurate. On the other hand, if $M(x)$ is large then $N \times y(x)$ is small, so the shape of the deflection curve is not very important.

The failure criterion for $N_\mathrm{b} = \chi N_\mathrm{o}$ is assumed such that the proof strength f_o is reached at the compressed extreme fibre in the section where $y = y_\mathrm{max}$. Then,

$$\frac{N_\mathrm{b}}{A} + \frac{N_\mathrm{b} y_\mathrm{max}}{W} = f_\mathrm{o} \tag{D6.27}$$

from which

$$y_\mathrm{max} = \frac{W f_\mathrm{o}}{N_\mathrm{b}} \left(1 - \frac{N_\mathrm{b}}{A f_\mathrm{o}}\right) \tag{D6.28}$$

The deflection curve for buckling of an elastic column in compression is a sine curve:

$$y(x) = y_\mathrm{max} \sin\left(\frac{\pi x}{l_\mathrm{cr}}\right) \tag{D6.29}$$

The stress for the beam column according to expression D6.26 can now be written as

$$\sigma(x) = \frac{N}{A} + \frac{M(x)}{W} + \frac{N f_\mathrm{o}}{N_\mathrm{b}} \left(1 - \frac{N_\mathrm{b}}{A f_\mathrm{o}}\right) \sin\left(\frac{\pi x}{l_\mathrm{cr}}\right) \tag{D6.30}$$

Failure for the beam column occurs for $\sigma(x) = f_\mathrm{o}$ in the most stressed section along the beam column. Insert $\sigma(x) = f_\mathrm{o}$ in expression D6.30, and divide the expression by f_o. Further, substitute the notations

$$N_\mathrm{o} = A f_\mathrm{o} \qquad M_\mathrm{o} = W f_\mathrm{o} \tag{D6.31}$$

The result is the following interaction formula:

$$\frac{N}{N_\mathrm{o}} + \left(\frac{N}{N_\mathrm{b}} - \frac{N}{N_\mathrm{o}}\right) \sin\left(\frac{\pi x}{l_\mathrm{cr}}\right) + \frac{M(x)}{M_\mathrm{o}} = 1 \tag{D6.32}$$

This equation is valid in the most stressed section, which is not necessarily the mid-section, as the moment $M(x)$ may be larger in other sections. In all other sections, the left hand side <1.

The second term in expression D6.32 is the influence of the second-order bending moment. If the term is multiplied by $M_\mathrm{o} = W f_\mathrm{o}$, the additional bending moment ΔM is found:

$$\Delta M = M_\mathrm{o} \left(\frac{N}{\chi N_\mathrm{o}} - \frac{N}{N_\mathrm{o}}\right) \sin\left(\frac{\pi x}{l_\mathrm{cr}}\right) = \frac{N W}{A} \left(\frac{1}{\chi} - 1\right) \sin\left(\frac{\pi x}{l_\mathrm{cr}}\right) \tag{D6.33}$$

This formula is given in clause 8.3 in Eurocode 3 (EN 1993-1-3, on cold-formed steel) for the design of splices and end connections, and should also be used for such sections in aluminium members, although it is not explicitly stated.

As $N_\mathrm{b} = \chi N_\mathrm{o}$, expression D6.32 can be recast as

$$\frac{N}{\chi N_\mathrm{o}} \left[\chi + (1 - \chi) \sin\left(\frac{\pi x}{l_\mathrm{cr}}\right)\right] + \frac{M(x)}{M_\mathrm{o}} = 1 \tag{D6.34}$$

or

$$\frac{N}{\omega_x \chi N_\mathrm{o}} + \frac{M(x)}{M_\mathrm{o}} \leq 1 \tag{D6.35}$$

Figure 6.38. Examples of K diagrams

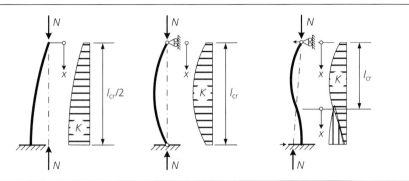

in which the following notation is inserted:

$$\omega_{\mathrm{x}} = \frac{1}{\chi + (1 - \chi)\sin(\pi x / l_{\mathrm{cr}})} \qquad \text{(D6.36)}$$

If the design section is in the middle of the beam column, then $x = l_{\mathrm{cr}}/2$, and $\omega_{\mathrm{x}} = 1$, and the denominator in expression D6.35 is the buckling resistance $\chi N_{\mathrm{o}} = N_{\mathrm{b}}$. If, on the other hand, the design section is at an end of the beam column, then $\omega_{\mathrm{x}} = 1/\chi$, and the denominator in expression D6.35 is the yield load N_{o}.

Except for the difference in notation and the exponents, expression D6.35 is the same as *expressions 6.59* and *6.60* in **clause 6.3.3.1**:

Clause 6.3.3.1

$$\left(\frac{N_{\mathrm{Ed}}}{N_{\mathrm{b,Rd}}}\right)^{\xi_{\mathrm{yc}}} + \frac{M_{\mathrm{y,Ed}}}{M_{\mathrm{y,Rd}}} \leq 1.00 \qquad \text{(D6.37)}$$

where $N_{\mathrm{b,Rd}} = \chi \omega_{\mathrm{x}} A f_{\mathrm{o}} / \gamma_{\mathrm{M1}}$ (if $\kappa = 1$ and $A_{\mathrm{eff}} = A$).

In design situations it may be practical to introduce short-hand notation for the terms in the interaction formulae:

$$K = \left(\frac{N_{\mathrm{Ed}}}{\chi N_{\mathrm{Rd}}}\left[\chi + (1 - \chi)\sin\left(\frac{\pi x}{l_{\mathrm{cr}}}\right)\right]\right)^{\xi_{\mathrm{yc}}} = \left(\frac{N_{\mathrm{Ed}}}{\omega_{\mathrm{x}}\chi N_{\mathrm{Rd}}}\right)^{\xi_{\mathrm{yc}}} \qquad \text{(D6.38)}$$

$$B = \frac{M_{\mathrm{Ed}}(x)}{M_{\mathrm{Rd}}} \qquad \text{(D6.39)}$$

Examples of K diagrams are given in Figure 6.38, where it can be seen that the K diagrams follow the deflection curves for elastic buckling. The design procedures for beam columns are illustrated in Examples 6.13 and 6.14 later in this guide, and the K-diagram was derived in Examples 6.11 for an axially loaded column with a stepwise variable cross-section.

6.3.3.2 Lateral torsional buckling

For members susceptible to torsional deformation, lateral-torsional buckling is often the decisive buckling mode. Example of cross-sections are open cross-sections symmetrical about the major axis (I cross-section) and centrally symmetric or double symmetric cross-sections (Z and I cross-sections). For such sections, the following criterion should be satisfied:

$$\left(\frac{N_{\mathrm{Ed}}}{N_{\mathrm{z,b,Rd}}}\right)^{\eta_{\mathrm{c}}} + \left(\frac{M_{\mathrm{y,Ed}}}{M_{\mathrm{y,b,Rd}}}\right)^{\gamma_{\mathrm{c}}} + \left(\frac{M_{\mathrm{z,Ed}}}{M_{\mathrm{z,Rd}}}\right)^{\xi_{\mathrm{zc}}} \leq 1.00 \qquad \text{(6.63)}$$

where the notation is the same as for flexural buckling (see Section 6.3.3.1) except that:

$M_{\mathrm{y,Ed}}$ is moment of the *first order* for beam-columns with hinged ends and members in non-sway frames, whereas for members in frames free to sway, $M_{\mathrm{y,Ed}}$ is bending moment according to *second order* theory.

Clause 6.3.1.1

$N_{z,b,Rd}$ is the axial force resistance for buckling in the x–y plane or torsional-flexural buckling according to *clause 6.3.1.1*.

Clause 6.3.2.1

$M_{y,b,Rd}$ is the bending moment resistance with respect to the y–y axis according to *clause 6.3.2.1*.

Clause 6.2.5.1

$M_{z,Rd}$ is the bending moment resistance with respect to the z–z axis according to *clause 6.2.5.1*.

The exponents are:

$$\eta_c = 0.8 \quad \text{or alternatively} \quad \eta_0 \chi_z \qquad \eta_c \geq 0.8$$

$$\gamma_c = \gamma_0$$

$$\xi_{zc} = 0.8 \quad \text{or alternatively} \quad \xi_0 \chi_z \qquad \text{but } \xi_{zc} = 0.8$$

Clause 6.2.9.1

where η_0, γ_0 and ξ_0 are defined in Section 6.2.9 (*clause 6.2.9.1*).

Clause 6.3.3.1

The criterion for flexural buckling (see Section 6.3.3.1 (*clause 6.3.3.1*)) should also be checked.

Example 6.14 (later in this chapter) illustrates the procedure for bi-axis bending.

6.3.3.3. Members containing localised welds

The influence on the buckling resistance of a local weakening in the HAZ around welds is dependent on the position of the weld along the member. Welds at the ends of a simply supported member have little influence on the buckling resistance of slender columns, as the second-order bending moment is zero at the ends. On the other hand, the HAZ weakening can be substantial if the weld is at the mid-section of the member. To allow for the HAZ weakening, the factors $\omega_{x,haz}$ and $\omega_{xLT,haz}$ are introduced in the resistance formulae for axial force and the bending moment.

Generally, the strength in the HAZ in a section with localised transverse welds should be based on the **ultimate** strength of the HAZ-softened material, if such softening occurs only locally along the length. It could be referred to the most unfavourable section in the bay considered. The reduction factor is then

$$\omega_{x,haz} = 1 \quad \text{and} \quad \omega_{xLT,haz} = 1 \tag{6.64}$$

However, if HAZ softening occurs close to the ends of the bay, or close to points of contra-flexure only, $\omega_{x,haz}$ and $\omega_{xLT,haz}$ may be increased when considering flexural and lateral torsional buckling, provided that such softening does not extend a distance along the member greater than the least width (e.g. flange width) of the section:

$$\omega_{x,haz} = \frac{1}{\chi_{haz} + (1 - \chi_{haz}) \sin(\pi x_{s,haz}/l_c)} \tag{6.65}$$

$$\omega_{xLT,haz} = \frac{1}{\chi_{LT,haz} + (1 - \chi_{LT,haz}) \sin(\pi x_{s,haz}/l_c)} \tag{6.66}$$

where:

$\chi_{haz} =$ $\chi_{y,haz}$ or $\chi_{z,haz}$, dependent on the buckling direction.

$\chi_{LT,haz}$ is the reduction factor for lateral torsional buckling of the beam column in bending only.

$x_{s,haz}$ is the distance from the localised weld to a support or point of contra-flexure for the deflection curve for elastic buckling of axial force only (compare Figure 6.40).

l_c is the buckling length.

Calculation of χ_{haz} ($\chi_{y,haz}$ or $\chi_{z,haz}$) and $\chi_{LT,haz}$ in the design section with the localised weld should be based on the ultimate strength of the heat-affected material for the slenderness

$$\bar{\lambda}_{haz} = \sqrt{\frac{A_{u,eff} f_u}{N_{cr}} \frac{\gamma_{M1}}{\gamma_{M2}}} = \bar{\lambda} \sqrt{\frac{A_{u,eff} f_u}{A_{eff} f_o} \frac{\gamma_{M1}}{\gamma_{M2}}} \tag{6.67}$$

$$\bar{\lambda}_{\text{LT,haz}} = \sqrt{\frac{W_{\text{u,eff}} f_{\text{u}}}{M_{\text{cr}}} \frac{\gamma_{\text{M1}}}{\gamma_{\text{M2}}}} = \bar{\lambda}_{\text{LT}} \sqrt{\frac{W_{\text{u,eff}} f_{\text{u}}}{\alpha W_{\text{el}} f_{\text{o}}} \frac{\gamma_{\text{M1}}}{\gamma_{\text{M2}}}} \qquad (6.68)$$

If the length of the softening region is larger than the least width (e.g. flange width) of the section, then the factor $\rho_{\text{u,haz}}$ for local failure in the expressions for $A_{\text{u,eff}}$ should be replaced by the factor $\rho_{\text{o,haz}}$ for overall yielding, leading to $A_{\text{u,eff}} = A_{\text{eff}}$, $\bar{\lambda}_{\text{haz}} = \bar{\lambda}$ and $\bar{\lambda}_{\text{LT,haz}} = \bar{\lambda}_{\text{LT}}$.

Examples 6.7, 6.10 and 6.11 show the influence of localised transverse HAZs.

6.3.3.4 Members containing a localised reduction of the cross-section

Members containing a localised reduction of the cross-section (e.g. unfilled bolt holes, oversized holes, slotted holes or flange cut-outs) may be checked according to Section 6.3.3.3 by replacing $\rho_{\text{u,haz}}$ with $A_{\text{net}}/A_{\text{g}}$, where A_{net} is a net cross-section area with a reduction for holes, and A_{g} is the gross cross-section area.

6.3.3.5 Design section of a member with unequal end moments

For members subjected to a combined axial force and unequal end moments and/or transverse loads, different sections along the beam column should be checked. If only end moments are present, then the design section can be found using *expression 6.71*. It is derived as follows.

If the notations

$$K_0 = \frac{N_{\text{Ed}}}{N_{\text{Rd}}} \qquad (D6.40a)$$

$$K_{\text{c}} = \frac{N_{\text{Ed}}}{\chi N_{\text{Rd}}} \qquad (D6.40b)$$

$$B_0 = \frac{M_{\text{Ed,1}}}{M_{\text{Rd}}} \qquad (D6.40c)$$

are introduced, the influence of the axial force according to expression 6.38 can be written as

$$K(x) = K_0 + (K_{\text{c}} - K_0) \sin\left(\frac{\pi x}{l_{\text{cr}}}\right) \qquad (D6.41)$$

For a linearly bending moment distribution with $M_{\text{Ed,1}}$ for $x = 0$ and $\psi M_{\text{Ed,1}}$ for $x = l_{\text{cr}}$,

$$B(x) = B_0 - (B_0 - \psi B_0)\frac{x}{l_{\text{cr}}} \qquad (D6.42)$$

The maximum of $K(x) + B(x)$ is found by derivation (see also Figure 6.39):

$$\frac{\text{d}}{\text{d}x}(K(x) + B(x)) = -\frac{\pi}{l_{\text{cr}}}(K_{\text{c}} - K_0)\cos\left(\frac{\pi x}{l_{\text{cr}}}\right) + (B_0 - \psi B_0)\frac{1}{l_{\text{cr}}} = 0 \qquad (D6.43)$$

Figure 6.39. Design section for linearly distributed bending moment

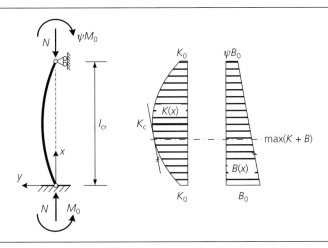

Figure 6.40. Buckling length l_c and definitions of x_s ($= x_A$ or x_B)

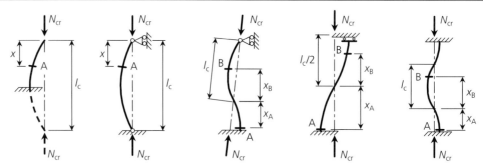

A and B are examples of studied sections marked with transverse lines.
See Figure 6.34 (*Table 6.8* in EN 1999-1-1) for values of the buckling length $l_c = kL$

from which

$$\cos\left(\frac{\pi x}{l_c}\right) = \frac{(B_0 - \psi B_0)}{\pi(K_c - K_0)} \qquad \text{but } x \geq 0 \text{ and } x \leq l_{cr} \tag{D6.44}$$

Clause 6.3.3.5

Inserting K_0, K_c, and B_0 according to expressions D6.40a–D6.40c, and substituting $M_{Ed,2} = \psi M_{Ed,1}$, we will obtain *expression 6.71* in **clause 6.3.3.5**:

$$\cos\left(\frac{x_s \pi}{l_c}\right) = \frac{(M_{Ed,1} - M_{Ed,2})}{M_{Rd}} \frac{N_{Rd}}{N_{Ed}} \frac{1}{\pi(1/\chi - 1)} \qquad \text{but } x_s \geq 0 \tag{6.71}$$

Examples of deflection curves and definitions of x_s for columns with different end conditions are shown in Figure 6.40.

Example 6.13: a member under major axis bending and compression

A beam column with a rectangular hollow section is loaded in major axis (y–y axis) bending according to Figure 6.41(a). The beam column is simply supported at A and fixed at B, and loaded with an axial force $N_{Ed} = 110$ kN and a concentrated load $F_{Ed} = 7.5$ kN. The material is EN AW-6063 T6 with proof strength $f_o = 160$ MPa.

The span is $L = 3.8$ m, and the load position is defined by $a = 0.2L$. The outer dimensions of the cross-section are $h = 120$ mm and $b = 80$ mm. The flange thickness is 5 mm, and the web thickness is 4 mm.

The beam column is assumed to be rigidly fixed at end B (not welded), so the theoretical buckling length $l_{cr} = 0.7L = 2.66$ m is used in this example, not the recommended value according to **clause 6.3.1.3**.

Clause 6.3.1.3

Calculation of the cross-section properties is not shown in this example. In axial compression, the cross-section is class 3, so the effective area is the same as the gross cross-section area $A = 1680$ mm^2. The cross-section resistance for an axial load is $N_{Rd} = 1680 \times 160/1.1 = 244$ kN, where 1.1 is the partial coefficient.

In y–y axis bending, the cross-section is class 2, so $M_{y,Rd}$ is the plastic moment resistance $= 10.2$ kN m, and the shape factor is $a_2 = 1.182$. The second moment of the area is $I_y = 3.534 \times 10^6$ mm^4.

For a rectangular cross-section, *expression 6.62* is valid. As $M_{z,Ed} = 0$, then the second term in the expression can be simplified (exponent $1.7 \times 0.6 \approx 1$), and the interaction formula will be

$$\left(\frac{N_{Ed}}{N_{b,Rd}}\right)^{\psi_c} + \frac{M_{y,Ed}}{M_{y,Rd}} \leq 1.00$$

where $N_{b,Rd} = \chi \omega_x N_{Rd}$.

Figure 6.41. Illustration of the design procedure for a beam column

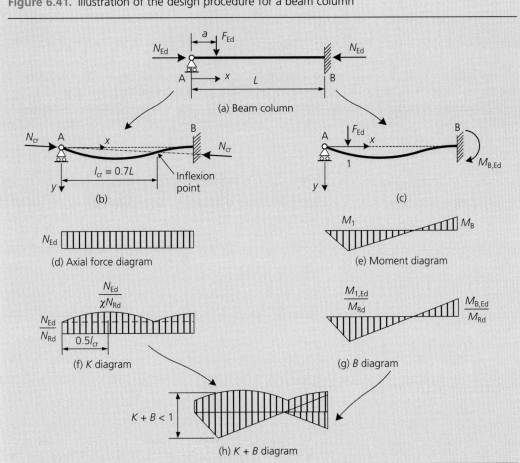

(a) Beam column

(b)

(c)

(d) Axial force diagram

(e) Moment diagram

(f) K diagram

(g) B diagram

(h) $K + B$ diagram

For the buckling length $l_{cr} = 0.7L = 2.66$ m, the elastic buckling load is

$$N_{cr} = \frac{\pi^2 EI_y}{l_{cr}^2} = \frac{\pi^2 \times 70\,000 \times 3.534 \times 10^6}{2660^2} = 345\,\text{kN}$$

and the slenderness is

$$\bar{\lambda} = \sqrt{\frac{Af_o}{N_{cr}}} = \sqrt{\frac{1680 \times 160}{345\,000}} = 0.883$$

which gives the reduction factor $\chi_y = 0.733$.

The exponent in the first term of the interaction formula is

$$\psi_c = 1.3\chi_y = 0.953$$

The K and B diagrams can now be sketched: see Figures 6.41(f) and 6.41(g).

As

$$\omega_x = \frac{1}{\chi_y + (1 - \chi_y)\sin(\pi x/l_{cr})}$$

then:

■   for $x = 0$ and $x = l_{cr}$,

$$\omega_x = \frac{1}{\chi_y}, \qquad K_0 = \left(\frac{N_{Ed}}{N_{Rd}}\right)^{\psi_c} = \left(\frac{110}{244}\right)^{0.953} = 0.468$$

- for $x = 0.5l_{cr}$,

$$\omega_x = 1, \qquad K_{0.5} = \left(\frac{N_{Ed}}{\chi_y N_{Rd}}\right)^{\psi_c} = \left(\frac{110}{0.733 \times 244}\right)^{0.953} = 0.629$$

As the shape of the K diagram is a sine curve, it can be sketched by hand according to Figure 6.41(f). Note that the K curve is mirrored at the inflexion point according to Figure 6.41(b).

For $x = L$,

$$\omega_x = \frac{1}{\chi_y + (1 - \chi_y)\sin(\pi x/l_{cr})} = \frac{1}{0.733 + (1 - 0.733)\sin(\pi \times 0.3 \times 3800/2660)} = 1.007$$

$$K_L = \left(\frac{E_{Ed}}{\omega_x \chi_y N_{Rd}}\right)^{\psi_c} = \left(\frac{110}{1.007 \times 0.733 \times 244}\right)^{0.953} = 0.625$$

just a little bit less than for $x = 0.5l_{cr}$.

The B curve is a scaled M curve.

For $x = a$

$$B_a = \frac{M_{1,Ed}}{M_{Rd}} = \frac{4.013}{10.21} = 0.393$$

For $x = L$

$$B_L = \frac{M_{B,Ed}}{M_{Rd}} = \frac{2.964}{10.21} = 0.290$$

The B diagram can now be drawn, and the two diagrams can be combined, as in Figure 6.41(h). Note, however, that for $x > l_{cr}$ the B curve is also mirrored around the abscissa to be able to add the K and B values.

We can see from the diagram that the maximum of $K + B$ occurs for $x = a$ where

$$\omega_x = \frac{1}{0.733 + (1 - 0.733)\sin(\pi \times 760/2660)} = 1.062$$

$$K_a = \left(\frac{110}{1.062 \times 0.733 \times 244}\right)^{0.953} = 0.594$$

$$K_a + B_a = 0.594 + 0.393 = 0.987 < 1.00$$

which is acceptable.

Example 6.14: lateral torsional buckling of a member in bi-axis bending and compression

The beam column in Figure 6.42 has an eccentric axial load, main axis eccentricity at one end and minor axis eccentricity at both ends. At the top, the load is applied via a rigid rectangular hollow section beam. It is simply supported at the load points A and D. In this example, lateral torsional buckling is checked according to *clause 6.3.3.2* (*expression 6.63*), and flexural buckling according to *clause 6.3.3.1* (*expression 6.59*).

Clause 6.3.3.2
Clause 6.3.3.1

Beam length	$l_{beam} = 2500$ mm	
Eccentricity at the top	$e_y = 400$ mm	$e_z = 30$ mm
Eccentricity at the bottom	$e_y = 0$ mm	$e_z = 30$ mm

Figure 6.42. General arrangement – loading and cross-section

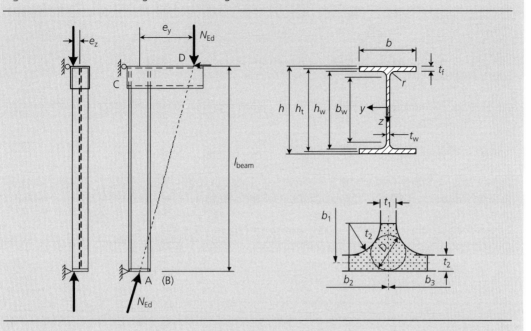

The cross-section dimensions are:

Section height $\quad\quad\quad\quad\quad\quad\quad\quad h = 200$ mm $\quad\quad\quad\quad$ flange width $b = 100$ mm
Web thickness $\quad\quad\quad\quad\quad\quad\quad t_w = 6$ mm $\quad\quad\quad\quad\quad$ flange thickness $t_f = 9$ mm
Fillet radius $\quad\quad\quad\quad\quad\quad\quad\quad r = 14$ mm
Web height $\quad\quad\quad\quad\quad\quad\quad\quad b_w = h - 2t_f - 2r = 154$ mm

EN AW-6082 T6 $\quad\quad\quad\quad\quad\quad f_o = 260$ MPa
Partial safety factor $\quad\quad\quad\quad\quad \gamma_{M1} = 1.1$

Axial compression force $\quad\quad\quad N_{Ed} = 60$ kN
Bending moment at end C $\quad\quad M_{y,Ed} = N_{Ed}e_y = 60 \times 0.4 = 24.0$ kN m
Bending moment at A and C $\quad M_{z,Ed} = N_{Ed}e_z = 60 \times 0.03 = 1.8$ kN m

Cross-section classification under axial compression (*clause 6.1.4*)

Clause 6.1.4

$$\varepsilon = \sqrt{250/f_o} = \sqrt{250/260} = 0.981$$

Outstand flanges (*expression 6.1*):

$$\beta_f = (b - t_w - 2r)/(2t_f) = (100 - 6 - 2 \times 14)/(2 \times 9) = 3.67$$

$$\beta_1 = 3\varepsilon = 2.94 < \beta_f$$

$$\beta_2 = 4.5\varepsilon = 4.41 > \beta_f$$

The flange is class 2.

Web – internal part in compression (*expression 6.1*):

$$\beta_w = b_w/t_w = 154/6 = 25.7$$

$$\beta_3 = 22\varepsilon = 21.6 < \beta_w$$

The web is class 4.

In compression, the overall cross-section classification is class 4. The resistance is therefore based on the effective cross-section for the member in compression.

Clause 6.1.4

Cross-section classification under *y–y* axis bending (*clause 6.1.4*)

The cross-section class for the compression flange is as for axial compression, so the flange is class 2.

The web is an internal part with a stress gradient that results in a neutral axis at the centre (*expression 6.2*):

$$\beta_w = 0.4b_w/t_w = 0.4 \times 154/6 = 10.3$$

$$\beta_1 = 11\varepsilon = 10.8 > \beta_w$$

The web is class 1.

In *y–y* axis bending, the overall cross-section classification is class 2. The resistance is therefore based on the plastic section modulus of the member.

Clause 6.1.4

Cross-section class under *z–z* axis bending (*clause 6.1.4*)

Outstand flanges (*expressions 6.4 and 6.3*):

$$\eta = 0.70 + 0.30\psi = 0.7 + 0.3 \times 0 = 0.7$$

$$\beta_f = \eta(b - t_w - 2r)/(2t_f) = 0.7(100 - 6 - 2 \times 14)/(2 \times 9) = 2.57$$

The limit for class 1 is 2.94 > 2.57, so the flange is class 1.

The web is in the neutral axis, so, in *z–z* axis bending, the overall cross-section classification is class 1. The resistance is based on the plastic section modulus of the member.

Clause 6.2.5

Design resistance for *y–y* axis bending (*clause 6.2.5*)

Although the resistance is based on the plastic section modulus, the elastic section modulus is needed to calculate the shape factors and the exponents in the interaction formulae. If the fillets are omitted, then

$$I_y = \frac{1}{12}\left(bh^3 - (b - t_w)(h - 2t_f)^3\right)$$

$$= \frac{1}{12}\left(100 \times 200^3 - (100 - 6)(200 - 2 \times 9)^3\right) = 1.94 \times 10^7 \, \text{mm}^4$$

The value, including fillets, can also be found in the CAD drawings:

$$I_y = 2.074 \times 10^7 \, \text{mm}^4$$

The elastic section modulus is

$$W_{el,y} = I_y/(h/2) = 2.074 \times 10^7/(200/2) = 2.074 \times 10^5 \, \text{mm}^3$$

Usually, the plastic section modulus cannot be obtained from the CAD program. Including fillets and using the notation (see Figure 6.42)

$$h_t = h - t_f = 200 - 9 = 191 \, \text{mm}$$

$$h_w = h - 2t_f = 200 - 2 \times 9 = 182 \, \text{mm}$$

we have

$$W_{pl,y} = bt_f h_t + \frac{1}{4}t_w h_w^2 + 2r^2(h_w - r) - \pi r^2\left[h_w - 2r\left(1 - \frac{4}{3\pi}\right)\right]\frac{1}{2}$$

$$= 100 \times 9 \times 191 + \frac{1}{4}(6 \times 182^2) + 2 \times 14^2(182 - 14) - \pi$$

$$\times 14^2\left[182 - 2 \times 14\left(1 - \frac{4}{3\pi}\right)\right]\frac{1}{2} = 2.364 \times 10^5 \, \text{mm}^3$$

The shape factor is

$$\alpha_y = W_{pl,y}/W_{el,y} = 2.364/2.074 = 1.140$$

and the resistance for y–y axis bending is

$$M_{y,Rd} = \alpha_y W_{el,y} f_o/\gamma_{M1} = 1.140 \times 2.074 \times 10^5 \times 260/1.1 = 55.9\,\text{kN m}$$

Design resistance for z–z axis bending (*clause 6.2.5*)

If the fillets are omitted, then (including fillets $I_z = 1.510 \times 10^6\,\text{mm}^4$)

$$I_z = \frac{1}{12}\left(2t_f b^3 + h_w t_w^2\right) = \frac{1}{12}\left(2 \times 9 \times 100^3 + 191 \times 6^3\right) = 1.503 \times 10^6\,\text{mm}^4$$

The elastic section modulus is

$$W_{el,z} = I_z/(b/2) = 1.510 \times 10^6/(100/2) = 3.020 \times 10^4\,\text{mm}^3$$

Although the influence of the fillets can be neglected, it is included here:

$$W_{pl,z} = \frac{1}{4}2t_f b^2 + \frac{1}{4}h_w t_w^2 + 2r^2(t_w + r) - \pi r^2\left[t_w + 2r\left(1 - \frac{4}{3\pi}\right)\right]\frac{1}{2}$$

$$= \frac{1}{2} \times 9 \times 100^2 + \frac{1}{4}(191 \times 6^2) + 2 \times 14^2(6 + 14) - \pi$$

$$\times 14^2\left[6 + 2 \times 14\left(1 - \frac{4}{3\pi}\right)\right]\frac{1}{2} = 4.767 \times 10^4\,\text{mm}^3$$

The shape factor and the resistance for z–z axis bending is

$$\alpha_z = W_{pl,z}/W_{el,z} = 4.767/3.020 = 1.578$$

$$M_{z,Rd} = f_o \alpha_z W_{el,z}/\gamma_{M1} = 260 \times 1.578 \times 3.020 \times 10^4/1.1 = 11.3\,\text{kN m}$$

Axial force resistance for y–y axis buckling (*clause 6.3.1*)

To calculate the effective cross-section area, the gross cross-section area is first calculated, and then the reduction due to local buckling is made:

$$A_{gr} = bh - (b - t_w)(h - 2t_f) + r^2(4 - \pi) = 100 \times 200 - 94 \times 182 + 14^2(4 - \pi) = 3060\,\text{mm}^2$$

The web slenderness according to the above is

$$\beta_w = b_w/t_w = 154/6 = 25.7$$

$$\beta_w/\varepsilon = 25.7/0.981 = 26.2$$

The reduction factor (*clause 6.1.5*) with $C_1 = 32$ and $C_2 = 220$ from Table 6.5 (*Table 6.3*) class A, no weld, is

$$\rho_c = \frac{C_1}{\beta/\varepsilon} - \frac{C_2}{(\beta/\varepsilon)^2} = \frac{32}{26.2} - \frac{220}{26.2^2} = 0.901 \tag{6.12}$$

$$A_{eff} = A_{gr} - b_w(t_w - \rho_c t_w) = 3060 - 154(6 - 0.901 \times 6) = 2969\,\text{mm}^2$$

The buckling length is $l_{cr,y} = 2500$ mm, so the buckling load and the slenderness will be

$$N_{cr,y} = \frac{\pi^2 E I_y}{l_{cr,y}^2} = \frac{\pi^2 \times 70\,000 \times 2.074 \times 10^7}{2500^2} = 2293\,\text{kN}$$

$$\bar{\lambda}_y = \sqrt{\frac{A_{eff} f_o}{N_{cr,y}}} = \sqrt{\frac{2969 \times 260}{2\,293\,000}} = 0.580 \tag{6.51}$$

Clause 6.2.5

Clause 6.3.1

Clause 6.1.5

The reduction factor for flexural buckling with $\alpha = 0.2$ and $\bar{\lambda}_0 = 0.1$ from *Table 6.6* for buckling class A is $\chi_y = 0.880$ (*expression 6.50*).

The buckling resistance, according to *expression 6.49* for $\kappa = 1$, no welds, is

$$N_{\mathrm{y,b,Rd}} = \kappa \chi_y f_o A_{\mathrm{eff}} / \gamma_{M1} = 1.0 \times 0.880 \times 260 \times 2969/1.1 = 618\,\mathrm{kN} \qquad (6.49)$$

The section resistance is needed in the interaction formulae:

$$N_{\mathrm{Rd}} = f_o A_{\mathrm{eff}} / \gamma_{M1} = 260 \times 2969/1.1 = 702\,\mathrm{kN}$$

Clause 6.3.1

Axial force resistance for *z–z* axis buckling (*clause 6.3.1*)
The buckling length is $l_{\mathrm{cr,y}} = 2520$ mm, so the buckling load and the slenderness will be

$$N_{\mathrm{cr,z}} = \frac{\pi^2 E I_z}{l_{\mathrm{cr,z}}^2} = \frac{\pi^2 \times 70\,000 \times 1.510 \times 10^6}{2500^2} = 167\,\mathrm{kN}$$

$$\bar{\lambda}_z = \sqrt{\frac{A_{\mathrm{eff}} f_o}{N_{\mathrm{cr,z}}}} = \sqrt{\frac{2969 \times 260}{167\,000}} = 2.150 \qquad (6.51)$$

The reduction factor for flexural buckling with $\alpha = 0.2$ and $\bar{\lambda}_0 = 0.1$ from *Table 6.6* for Buckling Class A is $\chi_z = 0.195$ (*6.50*).

Buckling resistance according to (*6.49*) for $\kappa = 1$ (no longitudinal welds) and $\omega_x = 1$

$$N_{\mathrm{z,b,Rd}} = \kappa \chi_z \omega_x A_{\mathrm{eff}} f_o / \gamma_{M1} = 1.0 \times 0.195 \times 1.0 \times 2969 \times 260/1.1 = 137\,\mathrm{kN} \qquad (6.49)$$

Clause 6.3.2.1

Lateral-torsional buckling of beam in bending (*clause 6.3.2.1*)
The elastic lateral-torsional buckling load is found in *Annex I*. You need the warping constant and the torsion constant found in *Annex J*. The warping constant is found in *Annex J, Figure J.2 case 5*.

$$I_w = (h - t_f)^2 I_z/4 = 191^2 \times 1.51 \times 10^6/4 = 1.377 \times 10^{10} \ \mathrm{mm}^6$$

The torsion constant including the fillets is

$$I_t = \sum bt^3/3 - 0.105 \sum t^4 + \sum aD^4 \qquad (J1.a)$$

where D is found in *Annex J, Figure J.1 case 2*

$$D = \left((\delta + 1)^2 + (\delta + 0.25 t_1/t_2) t_1/t_2 \right) \left(t_2/(2\delta + 1) \right)$$

$t_1 = t_w = 6$ mm and $t_2 = t_f = 9$ mm

$\delta = r/t_2 = 14/9 = 1.556$, see Figure 6.42

$\alpha = (0.10\delta + 0.15) t_1/t_2 = (0.10 \times 1.556 + 0.15)6/9 = 0.204$

$$D = \left((1.556 + 1)^2 + (1.556 + 0.25 \times 6/9)6/9 \right) \left(9/(2 \times 1.556 + 1) \right) = 16.8 \ \mathrm{mm}$$

Now I_t can be calculated for two flanges with fillets

$$I_t = \left(2 \times 100 \times 9^3 + 191 \times 6^3 \right)/3 - 0.105 \left(2 \times 9^4 + 6^4 \right) + 2 \times 0.204 \times 16.8^4$$

$$= 9.402 \times 10^4 \ \mathrm{mm}^4$$

(If the fillets are omitted then $I_t = 6.24 \times 10^4 \ \mathrm{mm}^4$).

The elastic critical moment for lateral-torsional buckling is given by the general formula

$$M_{cr} = \mu_{cr} \frac{\pi \sqrt{EI_z GI_t}}{L} \tag{I.2}$$

where the relative non-dimensional critical moment μ_{cr} is

$$\mu_{cr} = \frac{C_1}{k_z}\left[\sqrt{1 + \kappa_{wt}^2 + (C_2\zeta_g - C_3\zeta_j)^2} - (C_2\zeta_g - C_3\zeta_j)\right] \tag{I.3}$$

The standard conditions of restraint at each end are used which means $k_z = 1$, $k_w = 1$ and $k_y = 1$. (In reality the warping will be restrained at end C by the rectangular hollow section CD, omitted here for simplicity). The non-dimensional torsion parameter is then

$$\kappa_{wt} = \frac{\pi}{k_w L}\sqrt{\frac{EI_w}{GI_t}} = \frac{\pi}{1.0 \times 2500}\sqrt{\frac{70000 \times 1.377 \times 10^{10}}{27000 \times 9.402 \times 10^4}} = 0.774$$

As $k_z = 1$ the value of C_1 for any ratio of end moment loading (as indicated in *Table I.1*) is given approximately by

$$C_1 = (0.310 + 0.428\psi + 0.262\psi^2)^{-0.5} = (0.310 + 0 + 0)^{-0.5} = 1.796 \tag{I.6}$$

as $\psi = 0$. Values of C_2 and C_3 given in *Tables I.1* and *I.2*, are not needed in this case as $\zeta_g = 0$ and $\zeta_j = 0$.

The relative non-dimensional coordinate of the point of load application related to shear centre $\zeta_g = 0$ as well as the relative non-dimensional cross-section mono-symmetry parameter $\zeta_j = 0$. The formula for μ_{cr} is then simplified to

$$\mu_{cr} = \frac{C_1}{k_z}\left[\sqrt{1 + \kappa_{wt}^2}\right] = \frac{1.796}{1}\sqrt{1 + 0.774^2} = 2.272$$

Now the elastic critical moment for lateral-torsional buckling is found as

$$M_{cr} = \mu_{cr}\frac{\pi\sqrt{EI_z GI_t}}{L} = 2.272\frac{\pi\sqrt{70000 \times 1.510 \times 10^6 \times 27000 \times 9.402 \times 10^4}}{2500}$$

$$= 46.8 \text{ kNm} \tag{I.2}$$

The slenderness $\bar{\lambda}_{LT}$ is determined from

$$\bar{\lambda}_{LT} = \sqrt{\frac{\alpha W_{el,y}f_o}{M_{cr}}} \tag{6.58}$$

where α is taken from Table 6.8 (*Table 6.4*) subject to the limitation $\alpha \leq W_{pl,y}/W_{el,y}$. In this example $\alpha = \alpha_y = 1.140$.

$$\bar{\lambda}_{LT} = \sqrt{\frac{\alpha_y W_{el,y}f_o}{M_{cr}}} = \sqrt{\frac{1.140 \times 2.074 \times 10^5 \times 260}{46.8 \times 10^6}} = 1.146 \tag{6.58}$$

For class 1 and 2 cross-sections the parameters in the formulae for the reduction factor χ_{LT} for lateral torsional buckling are $\alpha_{LT} = 0.10$ and $\bar{\lambda}_{0,LT} = 0.6$ according to *paragraph 6.3.2.2(1)*. Then, for $\chi_{LT} = 0.675$ (6.56) and $\omega_{xLT} = 1$

Clause 6.3.2.2

$$M_{y,b,Rd} = \chi_{LT}\omega_{xLT}\alpha_y W_{el,y}f_o/\gamma_{M1} = 0.675 \times 1.0 \times 1.140 \times 2.074 \times 10^5 \times 260/1.1$$

$$= 37.7 \text{ kNm} \tag{6.55}$$

Clause 6.3.3.1
Clause 6.3.3.2

Interaction

Both flexural buckling according to *clause 6.3.3.1* and lateral-torsional buckling according to *clause 6.3.3.2* need to be checked, see *paragraph 6.3.3.2(2)*.

For major axis (y-axis) bending

$$\left(\frac{N_{Ed}}{N_{y,b,Rd}}\right)^{\xi_{yc}} + \frac{M_{y,Ed}}{M_{y,Rd}} \leq 1.00 \tag{6.59}$$

For lateral-torsional buckling

$$\left(\frac{N_{Ed}}{N_{z,b,Rd}}\right)^{\eta_c} + \left(\frac{M_{y,Ed}}{M_{y,b,Rd}}\right)^{\gamma_c} + \left(\frac{M_{z,Ed}}{M_{z,Rd}}\right)^{\xi_{zc}} \leq 1.00 \tag{6.63}$$

To shorten the formulae, the following notations are introduced:

$$K_0 = \frac{N_{Ed}}{N_{Rd}} = \frac{60}{702} = 0.0855$$

$$B_0 = \frac{M_{y,Ed}}{M_{y,Rd}} = \frac{24}{55.9} = 0.430$$

Clause 6.2.9.1(1)
Clause 6.3.1.1(1)
Clause 6.3.3.2(1)

The exponents η_0, γ_0 and ξ_0 in the interaction formulae for cross-section resistance are given in *clause 6.2.9.1(1)*. For flexural buckling, the exponents η_c, ξ_{yc} and ξ_{zc} are given in *clause 6.3.1.1(1)*, and for lateral torsional buckling the exponents η_c, γ_c and ξ_{zc} are given in *clause 6.3.3.2(1)*.

Conservatively, all exponents may be taken as 0.8. To illustrate the procedure, the expressions for the exponents are used:

$$\eta_0 = \alpha_z^2 \alpha_y^2 = 1.578^2 \times 1.140^2 = 3.24 \qquad \text{but } 1 \leq \eta_0 \leq 2 \rightarrow \eta_0 = 2$$

$$\gamma_0 = \alpha_z^2 = 1.578^2 \qquad \text{but } 1 \leq \gamma_0 \leq 1.56 \rightarrow \gamma_0 = 1.56$$

$$\xi_0 = \alpha_y^2 = 1.14^2 = 1.298 \qquad \text{but } 1 \leq \xi_0 \leq 1.56 \rightarrow \xi_0 = 1.298$$

$$\eta_c = \eta_0 \chi_z = 2 \times 0.195 = 0.390 \qquad \text{but } \eta_c \geq 0.8 \rightarrow \eta_c = 0.8$$

$$\xi_{yc} = \xi_0 \chi_y = 1.298 \times 0.88 = 1.143 \qquad \text{but } \xi_{yc} \geq 0.8 \rightarrow \xi_{yc} = 1.143$$

$$\xi_{zc} = \xi_0 \chi_z = 1.298 \times 0.195 \qquad \text{but } \xi_{zc} \geq 0.8 \rightarrow \xi_{zc} = 0.8$$

$$\gamma_c = \gamma_0 = 1.56$$

Clause 6.3.3.2
Clause 6.3.3.5(2)

Lateral torsional buckling check (*clause 6.3.3.2*)

The formula for defining the design section is, according to *clause 6.3.3.5(2)*,

$$\cos\left(\frac{x_s \pi}{l_c}\right) = \frac{(M_{Ed,1} - M_{Ed,2})}{M_{Rd}} \frac{N_{Rd}}{N_{Ed}} \frac{1}{\pi(1/\chi - 1)} \qquad \text{but } x_s \geq 0 \tag{6.71}$$

where $l_c = l_{cr,z}$, $M_{Ed,1} = M_{y,Ed}$, $M_{Ed,2} = \psi_y M_{y,Ed} = 0$ and the ratio between the moments at the ends is $\psi_y = 0$.

Expression 6.71 can now be evaluated:

$$\cos\left(\frac{x_s \pi}{l_{cr,z}}\right) = \frac{B_0}{K_0} \frac{1-0}{\pi(1/\chi_z - 1)} = \frac{0.430}{0.085} \frac{1}{\pi(1/0.195 - 1)} = 0.387 \qquad \text{but } x_s \geq 0$$

$$\frac{x_s\pi}{l_{cr,z}} = \text{acos}(0.387) = 1.173\,\text{rad}$$

$$x_s = 1.173 \times 2500/\pi = 934\,\text{mm}$$

The interaction expressions ω according to **clause 6.3.3.5(1)** are

Clause 6.3.3.5(1)

$$\omega_x = \frac{1}{\chi + (1-\chi)\sin(\pi x_s/l_{cr})} \qquad (6.69)$$

$$\omega_{xLT} = \frac{1}{\chi_{LT} + (1-\chi_{LT})\sin(\pi x_s/l_{cr})} \qquad (6.70)$$

The three terms in the interaction formula (*expression 6.63*) can be evaluated separately:

$$K_z = \left(\frac{K_0}{\chi_z \omega_x}\right)^{\eta_c} = \left[\frac{K_0}{\chi_z}\left(\chi_z + (1-\chi_z)\sin\frac{\pi x_s}{l_{cr,z}}\right)\right]^{\eta_c}$$

$$= \left[\frac{0.085}{0.195}\left(0.195 + (1-0.195)\sin\frac{\pi \times 934}{2500}\right)\right]^{0.8} = 0.491$$

$$B_y = \left(\frac{B_x}{\chi_{LT}\omega_{xLT}}\right)^{\gamma_c} = \left[\frac{B_0}{\chi_{LT}}\left(1-(1-\psi_y)\frac{x_s}{l_{cr,y}}\right)\left(\chi_{LT} + (1-\chi_{LT})\sin\frac{\pi x_s}{l_c}\right)\right]^{\gamma_c}$$

$$= \left[\frac{0.430}{0.675}\left(1-(1-0)\frac{934}{2500}\right)\left(0.675 - (1-0.675)\sin\frac{\pi \times 934}{2500}\right)\right]^{1.56} = 0.229$$

$$B_z = \left(\frac{M_{z,Ed}}{M_{z,Rd}}\right)^{\xi_{zc}} = \left(\frac{1.80}{11.3}\right)^{0.8} = 0.231$$

$$K_z + B_y + B_z = 0.491 + 0.229 + 0.231 = 0.951 < 1$$

The lateral torsional buckling check is acceptable.

For lateral torsional buckling, the design section is close to the centre of the beam due to a large second-order bending moment. This is illustrated in Figure 6.43. For flexural buckling, the second-order bending moment is small, so the design section will be at the top end.

Figure 6.43. *K* and *B* diagram and design sections (dash-dotted line). (a) Lateral torsional buckling. (b) Flexural buckling

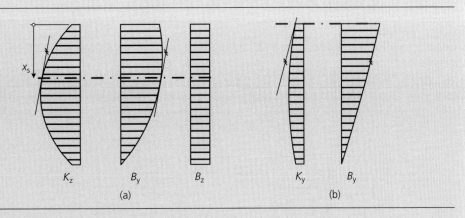

Flexural buckling check (*clause 6.3.3.1*)
Expression 6.71 for defining the design section according to **clause 6.3.3.5(2)** is now evaluated for y–y axis buckling:

Clause 6.3.3.1
Clause 6.3.3.5(2)

$$\cos\left(\frac{x_s\pi}{l_{cr,y}}\right) = \frac{B_0}{K_0}\frac{1-0}{\pi(1/\chi_y - 1)} = \frac{0.430}{0.085}\frac{1}{\pi(1/0.880 - 1)} = 11.7 \gg 1$$

There is no solution to this equation, which means that the design section is at the top end.

The two terms in the interaction formula (*expression 6.59*) can be evaluated separately:

$$K_y = \left[\frac{K_0}{\chi_y} \left(\chi_y + (1 - \chi_y) \sin \frac{\pi x_s}{l_{cr,y}} \right) \right]^{\xi_{yc}} = \left(\frac{0.085}{0.880} (0.880 + (1 - 0.880) \sin 0) \right)^{1.143} = 0.060$$

$$B_y = B_0 \left(1 - (1 - \psi_y) \frac{x_s}{l_{cr,y}} \right) = 0.430 \left(1 - (1 - 0) \frac{0}{2500} \right) = 0.430$$

$$K_y + B_y = 0.060 + 0.430 = 0.490 < 1$$

The flexural buckling check is acceptable.

6.4. Uniform built-up compression members

Clause 6.4

Clause 6.4 covers the design of uniform built-up compression members. The principal difference between the design of built-up columns and the design of conventional (solid) columns is in their response to shear. In conventional column buckling theory, lateral deflections are based on the flexural properties of the member, and the effects of shear on deflections are ignored. For built-up columns, shear deformations are far more significant due to the absence of a solid web, and therefore have to be accounted for in the development of design procedures.

There are two types of built-up member, laced and battened, characterised by the layout of the web elements, as shown in Figure 6.44. Laced columns contain diagonal web elements with or without additional horizontal web elements: these web elements are generally assumed to have pinned end conditions, and therefore to act in axial tension or compression. Battened columns (see Figure 6.44) contain horizontal web elements only and behave in the same manner as Vierendeel trusses, with the battens acting in flexure. Battened struts are generally more flexible in shear than laced struts.

Clause 6.4

Clause 6.4 also provides rules for closely spaced built-up members such as back-to-back channels. These rules will be commented on shortly in this guide.

In terms of material consumption, built-up members can offer much greater efficiency than single members. However, with the added expense of the fabrication process, the use of built-up members is not very popular, although there are special aluminium solutions where extruded profiles with long slots are expanded to a laced column.

6.4.1 General

Clause 6.4

Designing built-up members based on calculations of the discontinuous structure is considered too time-consuming for practical design purposes. *Clause 6.4* offers a simplified model that

Figure 6.44. Types of built-up compression members: (a) laced column; (b) battened column

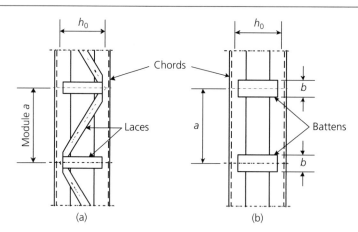

may be applied to uniform built-up compression members with pinned end conditions (although the code notes that appropriate modifications may be made for other end conditions). Essentially, the model replaces the discrete (discontinuous) elements of the built-up column with an equivalent continuous (solid) column, by 'smearing' the properties of the lacings or battens.

Design then comprises two steps:

1 Analyse the full 'equivalent' member with smeared shear stiffness using second order theory, as described in the following sub-section, to determine maximum design forces and moments.
2 Check critical chord and web members under design forces and moments. Joints must also be checked – see Chapter 8 of this guide.

The following rules regarding the application of the model are set out in ***clause 6.4.1***:   *Clause 6.4.1*

1 The chord members must be parallel.
2 The lacings or battens must form equal modules (i.e. uniform-sized lacings or battens and regular spacing).
3 The minimum number of modules in a member is three.
4 The method is applicable to built-up members with lacings in one or two directions, but is only recommended for members battened in one direction.
5 The chord members may be solid members or themselves built-up with lacings or battens in the perpendicular plane.

A bow imperfection of magnitude $L/500$ is employed in the design formulations of ***clauses***   *Clause 6.4.1(6)*
6.4.1(6) and ***6.4.1(7)***. The maximum design chord forces $N_{ch,Ed}$ are determined from the   *Clause 6.4.1(7)*
applied compression forces N_{Ed} and the applied bending moment M_{Ed}^{1}. The formulations were derived from the governing differential equation of a column and by considering second-order effects, resulting in the occurrence of the maximum design chord force at the mid-length of the column.

For a member with two identical chords, the design force $N_{ch,Ed}$ should be determined from

$$N_{ch,Ed} = 0.5 N_{Ed} + \frac{M_{Ed} h_0 A_{ch}}{2 I_{eff}} \qquad (6.72)$$

where

$$M_{ED} = \frac{N_{ED} e_0 + M_{ED}^{1}}{1 - \dfrac{N_{Ed}}{N_{cr}} - \dfrac{N_{Ed}}{S_v}}$$

$N_{cr} = \dfrac{\pi^2 E I_{eff}}{L^2}$ is the critical force of the effective built-up member

N_{Ed}	is the design value of the compression force to the built-up member
M_{Ed}	is the design value of the maximum moment in the middle of the built-up member considering second-order effects
M_{Ed}^{1}	is the design value of the maximum moment at the mid-length of the built-up member without second-order effects
h_0	is the distance between the centroids of chords
A_{ch}	is the cross-sectional area of one chord
I_{eff}	is the effective second moment of the area of the built-up member (see the following sections)
S_v	is the shear stiffness of the lacings or battened panel (see the following sections)
e_0	is the assumed imperfection magnitude and may be taken as $L/500$.

The lacings and the battens should be checked at the ends of the built-up member where the maximum shear occurs. The design shear force V_{Ed} should be taken as

$$V_{Ed} = \pi \frac{M_{Ed}}{L} \tag{6.73}$$

where M_{Ed} is as defined above.

6.4.2 Laced compression members

Clause 6.3.1
Clause 6.4.2.2

The chords and diagonal lacings of a built-up laced compression member should be checked for buckling in accordance with *clause 6.3.1*. Various recommendations on construction details for laced members are provided in *clause 6.4.2.2*.

Chords
The design compression force $N_{ch,Ed}$ in the chords is determined as described in the previous section. This should be shown to be less than the buckling resistance of the chords, based on a buckling length measured between the points of connection of the lacing system.

For lacings in one direction only, the buckling length of the chord L_{ch} may generally be taken as the system length. For lacings in two directions, buckling lengths are defined in the three-dimensional illustrations of *Figure 6.16*.

Lacings
The design compression force in the lacings may be determined from the design shear force V_{Ed} (described in the previous section) by joint equilibrium. Again, this design compressive force should be shown to be less than the buckling resistance. In general, the buckling length of the lacing may be taken as the system length.

Shear stiffness and effective second moment of area

Clause 6.4.2.1(3)
Clause 6.4.2.1(4)

The shear stiffness and effective second moment of area of the lacings required for the determination of the design forces in the chords and lacings are defined in *clauses 6.4.2.1(3)* and *6.4.2.1(4)*.

The shear stiffness S_v of the lacings depends upon the lacing layout, and, for the three common arrangements, reference should be made to *Figure 6.17*.

For laced built-up members, the effective second moment of area may be taken as

$$I_{eff} = 0.5h_0^2 A_{ch} \tag{6.75}$$

6.4.3 Battened compression members

Clause 6.4.3.2

The chords, battens and joints of battened compression members should be checked under the design forces and moments at mid-length and in an end panel. Various recommendations on design details for battened members are provided in *clause 6.4.3.2*.

Clause 6.4.3.1(2)

The shear stiffness S_v of a battened built-up member is given in *clause 6.4.3.1(2)*, and should be taken as

$$S_v = \frac{24EI_{ch}}{a^2\left(1 + \frac{2I_{ch}}{nI_b}\frac{h_0}{a}\right)} \leq \frac{2\pi^2 EI_{ch}}{a^2} \tag{6.76}$$

where

I_{ch} is the in-plane second moment of area of one chord about its own neutral axis
I_b is the in-plane second moment of area of one batten about its own neutral axis.

Clause 6.4.3.1(3)

The effective second moment of area I_{eff} of a battened built-up member is given in *clause 6.4.3.1(3)*, and may be taken as

$$I_{eff} = 0.5h_0^2 A_{ch} + 2\mu I_{ch} \tag{6.77}$$

where μ is a so-called efficiency factor, taken from *Table 6.9* of EN 1999-1-1. The second part of the right-hand side of *Equation 6.77*, $2\mu I_{ch}$, represents the contribution of the moments of inertia of the chords to the overall bending stiffness of the battened member. This contribution is not included for laced columns (see *Equation 6.75*). The primary reason behind this is that the spacing of the chords in battened built-up members is generally rather less than that for laced members, and it can therefore become uneconomical to neglect the chord contribution.

The efficiency factor μ, the value of which may range between zero and unity, controls the level of the chord contribution that may be exploited. It depends on the slenderness of the built-up member.

6.4.4   Closely spaced built-up members
Clause 6.4.4 covers the design of closely spaced built-up members. Essentially, provided the chords of the built-up members are either in direct contact with one another or closely spaced and connected through packing plates, and the conditions of *Table 6.10* of EN 1999-1-1 are met, the built-up members may be designed as integral members (ignoring shear deformations) following the provisions of *clause 6.3*; otherwise, the provisions of the earlier parts of *clause 6.4* apply.

Clause 6.4.4

Clause 6.3
Clause 6.4

6.5.   Unstiffened plates under in-plane loading
6.5.1   General
Clause 6.5 covers unstiffened plates as separate components under direct stress, shear stress or a combination of the two. The plates are attached to the supporting structure by welding, riveting, bolting or bonding, and the form of attachment can affect the boundary conditions. Thin plates must be checked for the ultimate limit states of bending under lateral loading, buckling under edge stresses in the plane of the plate, and for combinations of bending and buckling. The design rules in *clause 6.5* only refer to rectangular plates under in-plane loadings.

Clause 6.5

Clause 6.5

6.5.2   Resistance under uniform compression
Slender plates possess a significant post-critical resistance. For shorter plates with low aspect ratios a/b, this post-critical resistance gradually diminishes, because the 'two-dimensional' plate-like behaviour changes into 'one-dimensional' column-like behaviour that does not possess any post-critical resistance (Figure 6.45(c)).

For unstiffened panels this occurs at aspect ratios a/b well below 1.0, but for longitudinally stiffened panels with pronounced orthotropic properties, such behaviour may start at aspect ratios larger than $a/b = 1.0$ (see Figure 6.45(d)).

Figure 6.45. (a) Plate-like behaviour. (b) Column-like behaviour of an unstiffened plate with a small aspect ratio a/b. (c) Model of plate b. (d) Column-like behaviour of a longitudinally stiffened plate with aspect ratio $a/b > 1.0$

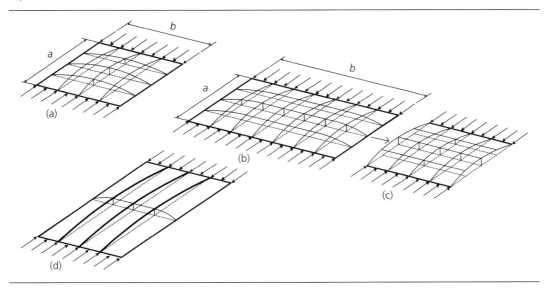

Clause 6.5.2(4) In *clause 6.5.2(4)* the reduction factor ρ_c is found from the more favourable of plate buckling and column buckling resistance, which means that there is no transition between the two, which in turn may lead to conservative results.

Plates in column-like buckling are treated as unsupported along the longitudinal edges, so the slenderness ratio, if the plate is simply supported at the loaded edges, is

$$\frac{l}{r} = \frac{a}{\sqrt{I/A}} = \frac{a}{\sqrt{t^3/12t}} \approx 3.5\frac{a}{t} \tag{D6.45}$$

For restrained loaded edges, a lower value of l/r can be used. The reduction factor χ for column buckling from Section 6.3.1 is used, based on the slenderness:

$$\bar{\lambda} = \sqrt{\frac{Af_o}{N_{cr}}} = \sqrt{\frac{Af_o}{\pi^2 EI/l^2}} = \frac{l}{\pi r}\sqrt{\frac{f_o}{E}} = 3.5\frac{a}{\pi t}\sqrt{\frac{f_o}{E}} \tag{D6.46}$$

Clause 6.1.5(2) For plate-like buckling, ρ_c is calculated from *clause 6.1.5(2)*, using the internal part expressions for plates that are simply supported, elastically restrained or fixed along longitudinal edges, and the outstand part expressions for plates with one longitudinal free edge. This means that restrained edges, if any, are not accounted for in plate buckling.

The classification, accounting for HAZ softening and holes, is the same as for members in Section 6.1.4.

6.5.3 Resistance under in-plane moment
If a pure in-plane moment acts on the ends of a rectangular unstiffened plate, the susceptibility to buckling is defined by the parameter $\beta = 0.4b/t$. The design procedure is the same as in Clause 6.2.5 appropriate parts of *clause 6.2.5* for members.

6.5.4 Resistance under a transverse or longitudinal stress gradient
If the applied actions at the end of a rectangular plate result in a transverse stress gradient, then Clause 6.5.2
Clause 6.5.3
Clause 6.5.6 the stresses are transferred into an axial force and a bending moment treated separately according to *clauses 6.5.2* and *6.5.3*. The load combination is then treated as in *clause 6.5.6*.

The yielding check should be performed at every cross-section, but for the buckling check a section at a distance equal to 0.4 times the elastic plate buckling half wavelength from the more heavily loaded end of the plate may be used.

6.5.5 Resistance under shear
The susceptibility to shear buckling is defined by the parameter $\beta = b/t$, where b is the shorter of the side dimensions. For all edge conditions, the plate in shear is classified as slender or non-slender with the limit $\beta = 39\varepsilon$ between the classes.

For non-slender plates ($\beta \le 39\varepsilon$):

$$V_{Rd} = A_{net}f_o/(\sqrt{3}\gamma_{M1}) \tag{6.88}$$

where A_{net} is the net effective area allowing for holes and HAZ softening. If the HAZ extends around the entire perimeter of the plate, the reduced thickness is assumed to extend over the entire cross-section. Holes may be ignored if their total cross-sectional area is less than 20% of the total cross-sectional area bt.

Clause 6.5.5 For slender plates ($\beta > 39\varepsilon$), the expressions in *clause 6.5.5* do not take advantage of tension field action, but if it is known that the edge supports for a plate are capable of sustaining a tension Clause 6.7.3 field, the treatment given in *clause 6.7.3* can be employed:

$$V_{Rd} = v_1 btf_o/(\sqrt{3}\gamma_{M1}) \tag{6.89}$$

where

$$v_1 = 17t\varepsilon\sqrt{k_\tau}/b \text{ but not more than } v_1 = k_\tau \frac{430t^2\varepsilon^2}{b^2}$$

$$k_\tau = 5.34 + 4.00(b/a)^2 \qquad \text{if } a/b \geq 1$$

$$k_\tau = 4.00 + 5.34(b/a)^2 \qquad \text{if } a/b < 1$$

6.5.6   Resistance under combined action

For combined axial force and in-plane moment, *condition 6.90* should be satisfied:

$$\frac{N_{Ed}}{N_{Rd}} + \frac{M_{Ed}}{M_{Rd}} \leq 1.00 \qquad\qquad (6.90)$$

If the combined action includes the effect of a coincident shear force, this shear force may be ignored if it does not exceed $0.5V_{Rd}$. If $V_{Ed} > 0.5V_{Rd}$, *condition 6.90b* should be satisfied:

$$\frac{N_{Ed}}{N_{Rd}} + \frac{M_{Ed}}{M_{Rd}} + \left(\frac{2V_{Ed}}{V_{Rd}} - 1\right)^2 \leq 1.00 \qquad\qquad (6.90b)$$

Example 6.15: resistance of an unstiffened plate under axial compression

The aim is to find the axial force resistance of a short rectangular hollow extrusion (Figure 6.46) made of aluminium.

Section properties

The cross-section dimensions are as follows: width $b = 200$ mm, thickness $t = 5$ mm. The inner dimension is then $b_i = b - 2t = 190$ mm. The material is EN AW-6082 T6 with $f_o = 260$ MPa. The partial safety factor $\gamma_{M1} = 1.1$. Length $a = 120$ mm.

Cross-section classification for axial compression (*clause 6.1.4*)

Clause 6.1.4

$$\varepsilon = \sqrt{250/f_o} = \sqrt{250/260} = 0.981$$

$$\beta = (b - 2t)/t = 190/5 = 38$$

For welds loaded at ends only, buckling class A, without welds, the limit for class 3 is

$$\beta_3 = 22\varepsilon = 21.6 < \beta$$

The classification is Class 4.

Reduction factor for plate-like buckling (*clause 6.1.5*)

Clause 6.1.5

Reduction factor for class 4 cross-section is

$$\rho_c = \min\left[C_1\frac{\varepsilon}{\beta} - C_2\left(\frac{\varepsilon}{\beta}\right)^2, 1.0\right] = 32\frac{0.981}{38} - 220\left(\frac{0.981}{38}\right)^2 = 0.679 \qquad (6.12)$$

Figure 6.46. Square cross-section

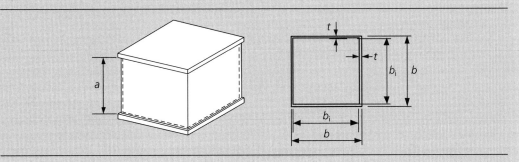

Clause 6.3.1

Reduction factor for column-like buckling (*clause 6.3.1*)
The plates are welded to thick end plates. The buckling length is then 0.5 times the length of the plate if fixed ends are assumed. Then, according to expression D6.46,

$$\bar{\lambda} = 3.5\frac{l_{cr}}{\pi t}\sqrt{\frac{f_o}{E}} = 3.5\frac{0.5 \times 30}{\pi \times 5}\sqrt{\frac{260}{70\,000}} = 0.815$$

The reduction factor for flexural buckling with $\alpha = 0.2$ and $\bar{\lambda}_0 = 0.1$ from *Table 6.6* for buckling class A is $\chi = 0.773$ (*expression 6.50*).

Axial force resistance for column-like buckling
The reduction factor is the largest of $\rho_c = 0.679$ and $\chi = 0.773$. The resistance is thus

$$N_{o,Rd} = \chi A f_o/\gamma_{M1} = 0.773 \times 3900 \times 260/1.1 = 713\,kN \qquad (6.80)$$

Axial force resistance in the HAZ
According to *Table 3.2b*, $f_{u,haz} = 185\,MPa$, so

$$N_{u,Rd} = A f_{u,haz}/\gamma_{M2} = 3900 \times 185/1.25 = 577\,kN \qquad (6.81)$$

Design axial force resistance

$$N_{Rd} = \min(N_{o,Rd}, N_{u,Rd}) = 577\,kN$$

6.6. Stiffened plates under in-plane loading

6.6.1 General

Clause 6.6

Clause 6.6 covers plates supported on all four edges reinforced with the following (Table 6.9 and Figure 6.47):

- one or two central or eccentric longitudinal stiffeners   *Clauses 6.6.2–6.6.5*
- three or more equally spaced longitudinal stiffeners or corrugations   *Clauses 6.6.2–6.6.5*
- orthotropic plating   *Clause 6.6.6*

Clause 6.1.4.3

Special rules applicable to extrusions with cross-section parts reinforced with one or two open stiffeners, symmetrically placed, are given in *clause 6.1.4.3*.

Table 6.9. Elastic support of stiffeners in stiffened plates

Stiffened plate		Stiffeners	Elastic support	*Expression*
1		One centric or eccentric stiffener	$c = \dfrac{0.27Et^3b}{b_1^2 b_2^2}$	6.97
2		Two symmetrical stiffeners	$c = \dfrac{1.1Et^3}{b_1^2(3b - 4b_1)}$	6.98
3		Multi-stiffened plate with open stiffeners	$c = \dfrac{8.9Et^3}{b^3}$	6.99
4		Multi-stiffened plate with closed or partly closed stiffeners	Orthotropic plate (see Section 6.6.6)	

Figure 6.47. Example of a stiffened plate

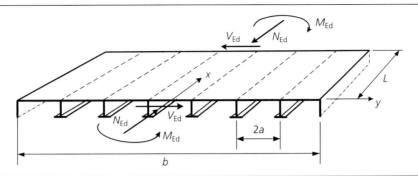

The stiffeners may be unsupported on their whole length or else be continuous over intermediate transverse stiffeners. The dimension L should be taken as the spacing between the supports. An essential feature of the design is that the longitudinal reinforcement, but not transverse stiffening, is 'sub-critical' (i.e. it can deform with the plating in an overall buckling mode).

The resistance of such plating to longitudinal direct stress in the direction of the reinforcement is given in *clauses 6.6.2* to *6.6.4*, and the resistance in shear is given in *clause 6.6.5*. Interaction between different effects may be allowed for in the same way as for unstiffened plates (see *clause 6.7.6*). The treatments are valid also if the cross-section contains class 4 cross-sections.

Clauses 6.6.2–6.6.4
Clause 6.6.5

Clause 6.7.6

If the structure consists of flat plating with longitudinal stiffeners, the resistance to transverse direct stress (e.g. patch loading) may be taken to be the same as for an unstiffened plate. With corrugated structures it is negligible. Orthotropic plating may have considerable resistance to transverse in-plane direct stress.

6.6.2   Stiffened plates under uniform compression
The resistance N_{Rd} for stiffened plates under uniform compression is the lesser of $N_{u,Rd}$ and $N_{c,Rd}$, where

$$N_{u,Rd} = A_{net} f_u / \gamma_{M2} \tag{6.92}$$

$$N_{c,Rd} = A_{eff} \chi f_o / \gamma_{M1} \tag{6.93}$$

A_{net} is the area of the least favourable cross-section, taking account of local buckling and HAZ softening if necessary, and also any unfilled holes.

A_{eff} is the effective area of the cross-section of the plating allowing for local buckling and HAZ softening due to longitudinal welds. HAZ softening due to welds at the loaded edges or at transverse stiffeners may be ignored in finding A_{eff}. Also, unfilled holes may be ignored. The plating is regarded as an assemblage of identical column sub-units, each containing one centrally located stiffener or corrugation and with a width equal to the pitch $2a$.

The reduction factor χ should be obtained from the appropriate column curve relevant to column buckling of the sub-unit as a simple strut out of the plane of the plating.

The slenderness $\bar{\lambda}$ in calculating χ is

$$\bar{\lambda}_c = \sqrt{\frac{A_{eff} f_o}{N_{cr}}} \tag{6.94}$$

where N_{cr} is the elastic orthotropic buckling load based on the gross cross-section.

For a plate with **open stiffeners**:

$$N_{cr} = \frac{\pi^2 EI_y}{L^2} + \frac{L^2 c}{\pi^2} \qquad \text{if } L < \pi \sqrt[4]{\frac{EI_y}{c}} \tag{6.95}$$

$$N_{cr} = 2\sqrt{cEI_y} \qquad \text{if } L \geq \pi \sqrt[4]{\frac{EI_y}{c}} \tag{6.96}$$

where c is the elastic support from the plate according to *expressions 6.97, 6.98* or *6.99* in Table 6.9 and I_y is the second moment of area of all stiffeners within the plate width b with respect to the y axes according to the figures in Table 6.9.

6.6.3   Stiffened plates under in-plane moment
The plating is regarded as an assemblage of column sub-units in the same general way as for axial compression.

6.6.4   Longitudinal stress gradient on multi-stiffened plates
For the column check it is sufficient to compare the design resistance with the design action effect arising at a distance $0.4l_w$ from the more heavily loaded end of a panel, where l_w is the half wavelength in elastic buckling according to *expression 6.100*:

$$l_w = \pi \sqrt[4]{\frac{EI_y}{c}} \tag{6.100}$$

where c is given in Table 6.9.

6.6.5   Multi-stiffened plating in shear

Clause 6.8.2

The design shear force resistance is found from **clause 6.8.2** (members with corrugated webs). Note, however, the difference in the coordinate system x and y for orthotropic plates and z and x for corrugated webs.

6.6.6   Buckling load for orthotropic plates
Examples of orthotropic plates are given in Table 6.10. Orthotropic plates can have one single layer (sections 1, 2 and 3) or double layers (sections 4, 5 and 6). Plates with one single layer have a large difference in the bending stiffness in the longitudinal and transverse directions, $B_y \ll B_x$.

The bending stiffness in the transverse direction B_y of section 2 is somewhat larger than the bending stiffness of a plane plate, thickness t, $B = Et^3/[12(1 - v^2)]$ (Figure 6.48(a)). The torsional stiffness is considerably larger than for a plane plate due to the closed parts. However, it is reduced due to cross-sectional distortion, illustrated in Figure 6.48(b).

For section 5, the distortion according to Figure 6.48(d) results in a reduced shear stiffness that is indirectly allowed for by reducing the transverse bending stiffness according to *expression 6.109a* (the quotient $10b^2/32a^2$); the effective torsional stiffness according to *expression 6.110* is also influenced.

Clause 6.6.2

For an orthotropic plate under **uniform compression** the procedure in **clause 6.6.2** may be used. The elastic orthotropic buckling load N_{cr} for a simply supported orthotropic plate is

Table 6.10. Orthotropic plates

No.	Open cross-sections	No.	Closed cross-sections
1		4	
2		5	
3		6	

Figure 6.48. Cross-section distortion. (a) Trapezoidal stiffeners in transverse bending, and (b) shear force. (c) Cell element in (d) shear or torsion

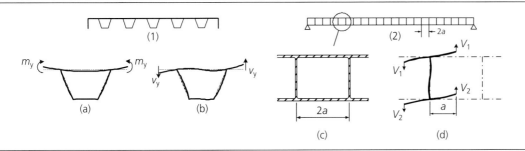

then given by

$$N_{cr} = \frac{\pi^2}{b}\left(\frac{B_x}{(L/b)^2} + 2H + B_y(L/b)^2\right) \qquad \text{if } \frac{L}{b} < \sqrt[4]{\frac{B_x}{B_y}} \tag{6.102}$$

$$N_{cr} = \frac{2\pi^2}{b}\left(\sqrt{B_x B_y} + H\right) \qquad \text{if } \frac{L}{b} \geq \sqrt[4]{\frac{B_x}{B_y}} \tag{6.103}$$

Expressions for B_x, B_y and H for different cross-sections are given in *Table 6.11* of EN 1999-1-1.

The extrusion technique makes it possible to produce double-skin plates (decks) by welding multi-hollow section profiles together with MIG or FSW (friction stir welding). Examples of profiles in vehicles and bridge decks are shown in Figure 6.49.

Figure 6.49. Examples of orthotropic plates: (a, b) for vehicles such as buses and trains; (c, d) for bridge decks

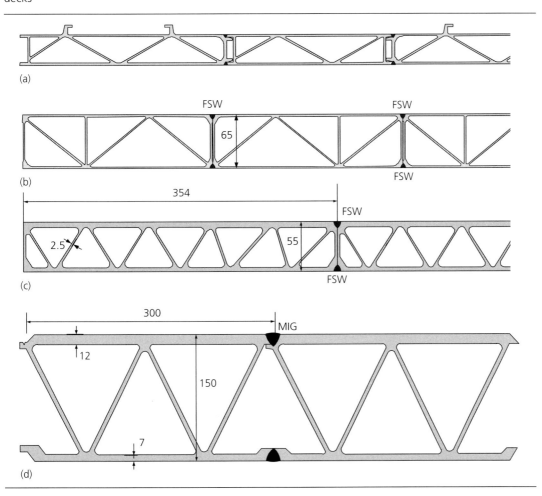

Section 6 in Table 6.10 consists of hollow section profiles put together with tongues and grooves. The plate has no transverse bending stiffness. However, the torsional stiffness is large, which has a significant influence on the resistance.

Example 6.16: resistance of an orthotropic plate under axial compression

An orthotropic plate is built up of hollow section extrusions welded together at the top and bottom layers (Figure 6.50). The plate forms the upper flange of a large box beam, and is loaded in axial compression and transverse load from traffic. In this example, the aim is to find the resistance for axial compression.

Section properties, plate dimensions and material

The cross-section dimensions are as follows: plate width $b = 2000$ mm, thickness of the upper layer $t_1 = 4$ mm, thickness of the bottom layer $t_2 = 3.6$ mm, thickness of the webs $t_3 = 3.4$ mm, half-width of the top layer $a_1 = 40$ mm and of the bottom layer $a_2 = 40$ mm. The depth of profile between mid-line of layers is $h = 60$ mm and width of webs $a_3 = 72.1$ mm. The length between rigid transverse stiffeners is $L = 5000$ mm.

The material is EN AW-6082 T6 with $f_o = 260$ MPa. The partial safety factor $\gamma_{M1} = 1.1$.

Clause 6.1.4

Cross-section classification for axial compression (*clause 6.1.4*)

$$\varepsilon = \sqrt{250/f_o} = \sqrt{250/260} = 0.981$$

The maximum slenderness is

$$\beta = (a_3 - 2r)/t_3 = (72.1 - 2 \times 5)/2.8 = 22.2$$

Longitudinal welds between extrusions are compensated for by an increase in material, and are found in every fourth node in the truss cross-section. Therefore, buckling class A, without welds, is used. The limit for class 3 is $\beta_3 = 22\varepsilon = 22.5 > \beta$. Thus, the classification is class 3, and there is no reduction due to local buckling.

Clause 6.6.6

Overall buckling, uniform compression (*clause 6.6.6*)

The cross-section area, centre of gravity and the second moment of area, omitting radii, are

$$A = 2t_1a_1 + 2t_2a_2 + 2t_3a_3 = 980 \, \text{mm}^2$$

$$e = (2t_2a_2h + 2t_3a_3h/2)/A = 28.0 \, \text{mm}$$

$$I_L = (2t_2a_2h^2 + 2t_3a_3h^2/3) - Ae^2 = 6.36 \times 10^5 \, \text{mm}^4$$

The torsion constant (from the handbook) is

$$I_t = \frac{4[h(a_1 + a_2)]^2}{2a_1/t_1 + 2a_2/t_2 + 2a_3/t_3} = 9.55 \times 10^5 \, \text{mm}^4$$

Figure 6.50. Orthotropic plate, cross-section

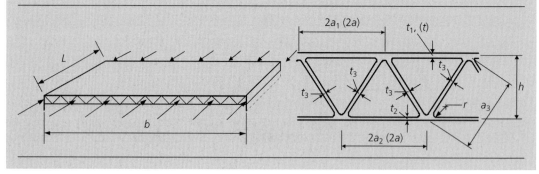

In *Table 6.11* in **clause 6.6.6**, the flexural and torsional rigidity of an orthotropic plate is given. For case 4:

$$B_x = \frac{EI_L}{2a} = 5.56 \times 10^8 \, \text{N mm}$$

$$B_y = \frac{Et_1 t_2 h^2}{t_1 + t_2} = 4.48 \times 10^8 \, \text{N mm}$$

$$H = \frac{GI_t}{2a} = 3.21 \times 10^8 \, \text{N mm}$$

Clause 6.6.6

The elastic buckling load is given by *expression 6.102* or *6.103*, depending on the plate length. In this case,

$$\frac{L}{b} = \frac{5000}{2000} = 2.5$$

$$\sqrt[4]{\frac{B_x}{B_y}} = \sqrt[4]{\frac{5.56}{4.48}} = 1.056$$

Thus, *expression 6.103* applies. Then

$$N_{cr} = \frac{2\pi^2}{b}\left(\sqrt{B_x B_y} + H\right) = 8099 \, \text{kN} \qquad (6.103)$$

Reduction factor for column-like buckling (*clause 6.6.2.4*)

Clause 6.6.2.4

The elastic buckling load corresponds to the total area of the orthotropic plate. The corresponding cross-section and the slenderness are

$$A_{eff} = A\frac{b}{2a} = 24\,500 \, \text{mm}^2 \qquad (6.50)$$

$$\bar{\lambda} = \sqrt{\frac{A_{eff} f_o}{N_{cr}}} = 0.852$$

The reduction factor for flexural buckling with $\alpha = 0.2$ and $\bar{\lambda}_0 = 0.1$ from *Table 6.6* in **clause 6.3.1.2** for buckling class A is $\chi = 0.751$ (*expression 6.50*).

Clause 6.3.1.2

Axial force resistance for column-like buckling (*clause 6.6.2(3)*)

Clause 6.6.2(3)

The reduction factor is $\chi = 0.751$, so the resistance is

$$N_{o,Rd} = \chi A f_o / \gamma_{M1} = 0.751 \times 24\,500 \times 240/1.1 = 4016 \, \text{kN} \qquad (6.93)$$

Alternative cross-sections

Alternative cross-sections are compared in Table 6.11. The thickness of the upper and bottom layers and the webs are the same for the first three cases, which means that the cross-section areas are not the same. However, the stresses for the axial force resistances indicate the efficiency of the sections. Case 1 is the example above. The calculations for the rest are not shown.

Case 1 with a truss cross-section can carry the largest axial load. Case 2 with profiles joined with groove and tongue is almost as efficient, although the transverse bending stiffness in the joints is zero. Case 3 with transverse webs is less efficient. Case 4, a plane plate with a uniform thickness giving the same cross-section area as case 1, can evidently carry considerably less load, although the post-critical resistance is utilised according to **clause 6.5.2(4)a**. If post-critical resistance is not utilised (Case 5, using **clause 6.5.2(4)b**)), then the resistance is very small for such a slender plate ($b/t = 163$).

Clause 6.5.2(4)a
Clause 6.5.2(4)b

Table 6.11. Resistance of alternative cross-sections

Case	Cross-section	Axial force resistance: kN	Gross cross-section area: A_g	Stress based on A_g: MPa
1		4016	24 500	164
2	Profiles joined with groove and tongue	3410	21 300	160
3		1580	18 600	85
4	Post-buckling strength utilised	1023	24 500	42
5	Post-buckling strength not utilised	400	24 500	8.2

6.7. Plate girders

6.7.1 General

A plate girder is a deep beam with a web that is usually slender and may therefore be reinforced by intermediate transverse stiffeners and/or longitudinal stiffeners. The web may buckle in shear at relatively low applied loads, but a considerable amount of post-buckled strength can be mobilised due to tension field action. Plate girders are sometimes constructed with transverse web reinforcement in the form of corrugations or closely spaced transverse stiffeners.

Plate girders can be subjected to combinations of moment, shear and axial loading, and to local loading on the flanges. Because of their slender proportions they may be subjected to lateral torsional buckling, unless properly supported along their length.

The rules for plate girders given in this section are generally applicable to the webs of box girders.

Actions on plate girder structures and corresponding resistance are given in the following clauses:

- Bending of girders with transverse web stiffeners only   *Clause 6.7.2*
- Bending of girders with longitudinal web stiffeners   *Clause 6.7.3*
- Bending of members with corrugated webs   *Clause 6.8.1*
- Shear buckling of plain webs   *Clause 6.7.4*
- Shear buckling of corrugated webs   *Clause 6.8.2*
- Interaction between shear force and bending moment   *Clause 6.7.6*
- Buckling of web due to local loading on flanges   *Clause 6.7.5*
- Flange induced web buckling   *Clause 6.7.7*
- Lateral torsional buckling   *Clause 6.3.2*
- Influence of shear lag   *Annex K*

6.7.2 Resistance of girders under in-plane bending

Clause 6.7.2

Clause 6.7.2 covers the design of girders with transverse web stiffeners. For webs with continuous longitudinal welds the effect of the HAZ should be investigated: however, the HAZ effect caused by the welding of transverse stiffeners may be neglected and small holes in the web may be ignored provided they do not occupy more than 20% of the cross-sectional area of the web.

A yielding check and a buckling check should be made. For the yielding check, the design value of the moment, M_{Ed} at each cross-section shall satisfy

$$M_{Ed} \leq M_{o,Rd} \tag{6.115}$$

where $M_{o,Rd}$, for any class cross-section, is the design moment resistance of the cross-section that would apply if the section were designated class 3. Thus,

$$M_{o,Rd} = W_{net} f_o / \gamma_{M1} \qquad\qquad (6.116)$$

where W_{net} is the elastic modulus allowing for holes and taking a reduced thickness $\rho_{o,haz}t$ in regions adjacent to the flanges that might be affected by HAZ softening (see ***clause 6.1.6.2***).

Clause 6.1.6.2

For the buckling check, the design moment resistance $M_{o,Rd}$ is given by

$$M_{o,Rd} = W_{eff} f_o / \gamma_{M1} \qquad\qquad (6.117)$$

where W_{eff} is the effective elastic modulus obtained by taking a reduced thickness to allow for local buckling as well as HAZ softening, but with the presence of holes ignored. The reduced thickness is equal to the lesser of $\rho_{o,haz}t$ and $\rho_c t$ in HAZ regions, and $\rho_c t$ elsewhere (see ***clause 6.2.5***).

Clause 6.2.5

In applying the buckling check, it is assumed that the spacing between adjacent transverse stiffeners is greater than half of the clear depth of the web between flange plates. If this is not the case, refer to ***clause 6.8*** for corrugated or closely stiffened webs.

Clause 6.8

The buckling check resembles ordinary cross-section checks, but actually relates to the whole panel length a between adjacent transverse stiffeners (buckling verification on the panel length). For this reason, the plate buckling verification may be performed at a distance $0.4a$ or $0.5h_w$, whichever is smaller, from the panel end with larger stresses, where h_w is the web depth between flanges. When large moment gradients are present, this may be very favourable. In this case, the additional cross-section check has to be performed at the end of the panel. Conservatively, the buckling check may be performed at the most stressed end of the panel. If there are transverse stiffeners at the plate girder ends only, the buckling check should be performed for the maximum bending moment.

The thickness is reduced in any class 4 part that is wholly or partly in compression (b_c in Figure 6.51). The stress ratio ψ used in ***clause 6.1.4.3*** and the corresponding width b_c may be obtained using the effective area of the compression flange and the gross area of the web (see Figure 6.51(c), gravity centre G_1). Thus, iteration is not needed in this case.

Clause 6.1.4.3

If the compression edge of the web is nearer to the neutral axis of the girder than in the tension flange (see Figure 6.51(c)), the method in ***clause 6.1.4.3*** may be used. This procedure generally requires an iterative calculation in which ψ is determined again at each step from the stresses calculated on the effective cross-section defined at the end of the previous step. See Section 6.2.5 in this guide.

Clause 6.1.4.3

Figure 6.51. Plate girder in bending. (a) Cross-section notation. (b) Effective cross-section for a symmetric plate girder with class 1, 2 and 3 flanges. (c) Effective cross-section for a girder with a smaller tension (bottom) flange and a class 4 compression (top) flange. (Reproduced from EN 1999-1-1 (*Figure 6.25*), with permission from BSI)

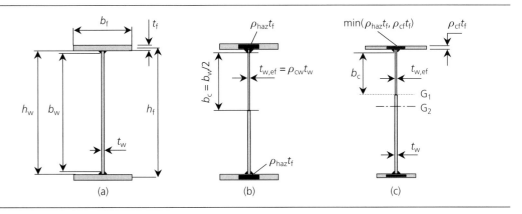

Figure 6.52. Stiffened web of a plate girder in bending. (a) Stiffened web. (b) Cross-section. (c) Effective area of the stiffener column. (d1, d2) Column cross-section for calculation of I_{st}. (Reproduced from EN 1999-1-1 (*Figure 6.26*), with permission from BSI)

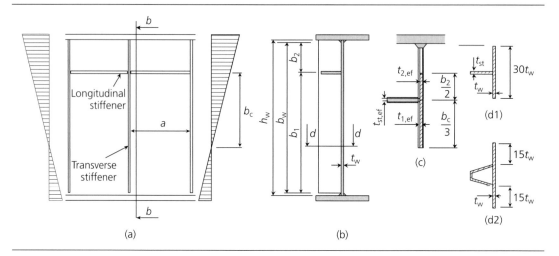

6.7.3 Resistance of girders with longitudinal web stiffeners

Clause 6.7.3

Clause 6.7.3 covers the design of plate girders with longitudinal web stiffeners. Local (plate) buckling due to longitudinal compressive stresses is taken into account by the use of an effective cross-section applicable to class 4 cross-sections, based on the effective thickness of the compression parts of the cross-section.

The overall plate buckling, including buckling of the stiffeners (also denoted distortional buckling), is considered as flexural buckling of an equivalent column consisting of the stiffeners and half of the adjacent parts of the web. If the stresses change from compression to tension within the sub-panel, one-third of the compressed part is taken as part of the column. See Figure 6.52(c).

The calculation is made in two steps:

1 The effective areas of flat compression sub panels between stiffeners are obtained using effective thicknesses according to Section 6.1.5 (see Figure 6.52).
2 The effective thickness of the different parts of the equivalent column section are further reduced with a reduction factor χ obtained from the appropriate column curve relevant for column buckling of the column as a simple strut out of the plane of the web.

The slenderness $\bar{\lambda}$ in calculating χ is

$$\bar{\lambda} = \sqrt{\frac{A_{st,eff}f_o}{N_{cr}}} \qquad (6.118)$$

where:

$A_{st,eff}$ is the effective area of the column from the first step, see Figure 6.52c
N_{cr} is the elastic buckling load given by

$$N_{cr} = 1.05E\frac{\sqrt{I_{st}t_w^3 b_w}}{b_1 b_2} \qquad \text{if } a > a_c \qquad (6.119)$$

$$N_{cr} = \frac{\pi^2 EI_{st}}{a^2} + \frac{Et_w^3 b_w a^2}{4\pi^2(1-v^2)b_1^2 b_2^2} \qquad \text{if } a \le a_c \qquad (6.120)$$

$$a_c = 4.33\sqrt[4]{\frac{I_{st}b_2^2 b_1^2 b_2^2}{t_w^3 b_w}} \qquad (6.121)$$

Figure 6.53. State of stress and the collapse behaviour of a plate girder subjected to shear. (a) Pure shear stress. (b) Tension field action, $\sigma_1 > \sigma_2$. (c) Equilibrium. (d) Plastic flange mechanism

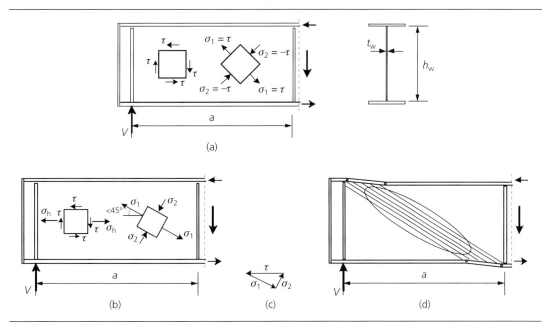

I_{st} is second moment of area of the gross cross-section of the stiffener and adjacent part of web about an axis through its centroid and parallel to the plane of the web

b_1, b_2 are distances from longitudinal edges to the stiffener ($b_1 + b_2 = b_w$)

a_c is the half wave length for elastic buckling of stiffener.

For calculation of I_{st}, the column consists of the actual stiffener together with an effective width $15t_w$ of the web plate on both sides of the stiffener (see Figures 6.52(d1) and 6.52(d2)), even if this effective width is larger than b_2.

In the case of two longitudinal stiffeners, both in compression, the two stiffeners are considered as lumped together, with an effective area and a second moment of area equal to the sum of those of the individual stiffeners. The location of the lumped stiffener is the position of the resultant of the axial forces in the stiffeners. If one of the stiffeners is in tension, the procedure will be conservative.

6.7.4   Resistance to shear

Behaviour

Plate girder webs are usually so slender that the resistance is influenced by shear buckling. Prior to buckling, pure shear stresses occur in the plate. If these shear stresses τ are transferred into principal stresses, they correspond to principal tensile stresses σ_1 and principal compression stresses σ_2 with equal magnitude and inclined by $45°$ with regard to the longitudinal axis of the girder. See Figure 6.53(a).

The compressive stresses in the $45°$ direction may be interpreted as the cause of the buckling of the plate. The shape of the buckles will be more complicated than in the case of, for example, uniform compression. An example from Timoshenko and Gere (1961) is given in Figure 6.54(a).

The critical shear buckling stress is given by expression D6.47 and the first terms of *expression 6.128*

$$\tau_{cr} = k_\tau \frac{\pi^2 E}{12(1 - v^2)} \left(\frac{t_w}{h_w}\right)^2 \tag{D6.47}$$

$$k_\tau = 5.34 + \frac{4}{(a/h_w)^2} \qquad \text{for } a \geq h_w \tag{part of 6.128}$$

Figure 6.54. Shear buckling of long rectangular plates. (a) Initial buckling pattern (Timoshenko and Gere, 1961). (b) Buckling pattern at load level $2.9\tau_{cr}$ (Bergman, 1948)

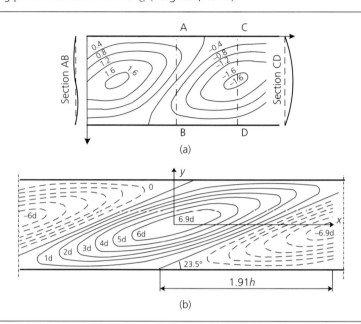

The buckling coefficient according to *expression 6.128* is not exact but rather a generally accepted approximation. If $a < h_w$, the solution can still be used, but with interchanged symbols for the panel sides.

As for plates subjected to direct compressive stress, slender plates under shear possess a post-critical reserve. After buckling, the plate reaches the post-buckling state of stress, while a shear buckle forms in the direction of the principal tensile stresses σ_1. Due to buckling, no significant increase in the stresses in the direction of the principal compressive stresses σ_2 is possible, whereas the principal tensile stresses can still increase. As a result, stress values of different magnitude occur (tension > compression), which lead to a rotation of the stress field for equilibrium reasons (see Figure 6.53(c)). This is denoted tension field action (see Figure 6.53(b)).

The development of such a tensile force is only possible if the boundary elements provide a sufficient anchorage for the axial forces. However, it can be shown that intermediate transverse stiffeners are not needed (see Figure 6.54(b)) for the development of the tension field, but they are practically always needed at the supports. The maximum amount of axial force that can be carried depends on the stiffness of the end post and the flexural rigidity of the flanges. Since the flanges restrain the relative deformation of the transverse stiffener to each other, the tension field can be anchored also in parts of the flanges. When reaching ultimate load, a plastic hinge mechanism forms in the flanges (see Figure 6.53(d)).

Clause 6.7.4

Design according to EN 1999-1-1 (*clause 6.7.4*)

There are many tension field theories that aim to describe the ultimate resistance of plate girders under shear (Figure 6.55). In EN 1999-1-1 (as in EN 1993-1-5 for steel) the rotated stress field theory proposed by Höglund (1971, 1973) and further developed in Höglund (1995) was adopted. Originally, it was developed for girders with web stiffeners at the supports only (e.g. webs with large aspect ratios ($a/h_w > 3$)) because other existing models led to very conservative results in this case. In EN 1999-1-1 and EN 1993-1-5 the rotated stress field theory was generally accepted, since it provides adequate results regardless of the panel aspect ratio. Furthermore, it could be used for longitudinally stiffened webs as well.

In this method, the shear resistance V_{Rd} comprises contributions from the web $V_{w,Rd}$ and from the flanges $V_{f,Rd}$ according to *expression 6.124*:

$$V_{Rd} = V_{w,Rd} + V_{f,Rd} \tag{6.124}$$

Figure 6.55. Examples of different tension field theories (Höglund, 1973; see therein for cited references)

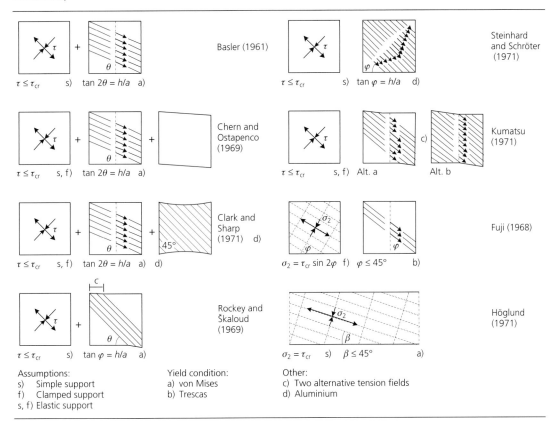

Assumptions:
s)   Simple support
f)   Clamped support
s, f) Elastic support

Yield condition:
a)   von Mises
b)   Trescas

Other:
c)   Two alternative tension fields
d)   Aluminium

where:

$V_{w,Rd}$ is the resistance from the rotated stress field in the web according to *expression 6.125* in *clause 6.7.4.2(4)*

$V_{f,Rd}$ is the resistance of a tension field anchored in the flanges according to *expression 6.131* in *clause 6.7.4.2(10)*.

Clause 6.7.4.2(4)

Clause 6.7.4.2(10)

The contribution from the web to the design resistance for shear should be taken as

$$V_{w,Rd} = \rho_v t_w h_w \frac{f_o}{\sqrt{3}\gamma_{M1}} \qquad (6.125)$$

where ρ_v is the factor for shear buckling obtained from expressions D6.48 and D6.49 (from *Table 6.13*) or Figure 6.56.

Reduction factor ρ_v

The reduction factor ρ_v considers components of pure shear and the anchorage of membrane forces by transverse stiffeners due to tension field action. Since the axial and flexural stiffness of the transverse end stiffeners influences the post-critical reserve, expressions D6.48 and D6.49 distinguish between rigid and non-rigid end posts in the determination of the reduction factor ρ_v (see Figure 6.56). Requirements for rigid end posts are given in *clause 6.7.8.1*. Note that a rigid end post may be assumed at an inner support of continuous girders even if there is only one double-sided stiffener above the support. See *clause 6.7.4.1*. The reason for this is that the longitudinal membrane stresses are balanced on the two sides of the support.

Clause 6.7.8.1

Clause 6.7.4.1

For a rigid end post:

$$\rho_v = \eta \qquad \text{if } \lambda_w \leq 0.83/\eta \qquad (D6.48a)$$

$$\rho_v = 0.83/\lambda_w \qquad \text{if } 0.83/\eta < \lambda_w < 0.937 \qquad (D6.48b)$$

$$\rho_v = 2.3/(1.66 - \lambda_w) \qquad \text{if } \lambda_w \geq 0.937 \qquad (D6.48c)$$

Figure 6.56. Factor ρ_v for shear buckling and the critical curve $1/\lambda_w^2$

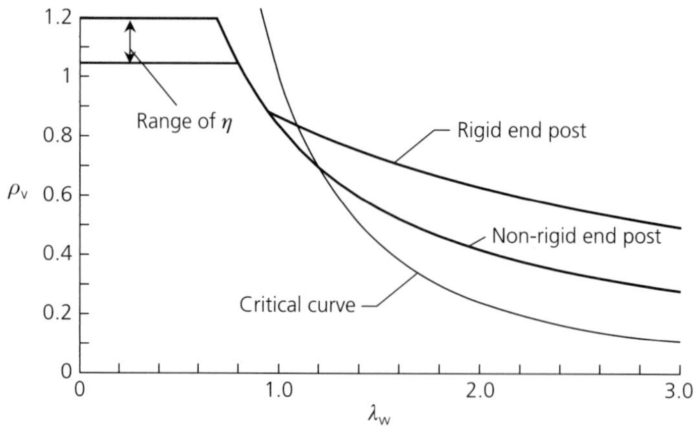

For a non-rigid end post:

$$\rho_v = \eta \qquad \text{if } \lambda_w \leq 0.83/\eta \tag{D6.49a}$$

$$\rho_v = 0.83/\lambda_w \qquad \text{if } \lambda_w > 0.83/\eta \tag{D6.49b}$$

In the expressions

$$\eta = 0.7 + 0.35 f_{uw}/f_{ow} \qquad \text{but not more than 1.2} \tag{D6.50}$$

where f_{ow} is the strength for overall yielding and f_{uw} is the ultimate strength of the web material.

As illustrated in Figure 6.56, $\rho_v = \eta$ is defined for small slenderness with an η value larger than 1 according to expression D6.50. The reason for this is that strain hardening in shear can be tolerated in this case since it does not lead to excessive deformations. The shear stress is then larger than corresponding to initial yielding, $f_o/\sqrt{3} \approx 0.58 f_o$.

The reduction curves according to expressions D6.48 and D6.49 apply for the verification of both unstiffened and stiffened webs. They are based on the plate slenderness λ_w.

Slenderness λ_w

In many cases, intermediate stiffeners are not needed, just stiffeners at the ends of the plate girder. Then, the distance between the transverse stiffeners is large, and $k_\tau \cong 5.34$. The slenderness is then

$$\lambda_w = \sqrt{\frac{f_o/\sqrt{3}}{\tau_{cr}}} = \sqrt{\frac{f_o/\sqrt{3}}{5.34 \dfrac{\pi^2 E}{12(1-v^2)} \dfrac{t_w^2}{b_w^2}}} = 0.35 \frac{b_w}{t_w} \sqrt{\frac{f_o}{E}} \qquad \text{for } v = 0.3 \tag{6.123}$$

In the case of a stiffened panel, the largest value of the slenderness of the sub-panels and the stiffened panel governs. The slenderness λ_w is determined according to

$$\lambda_w = \frac{0.81}{\sqrt{k_\tau}} \frac{b_w}{t_w} \sqrt{\frac{f_o}{E}} \tag{6.126}$$

Clause 6.7.8.3

in which k_τ is the minimum shear buckling coefficient for the web panel. Rigid boundaries may be assumed if flanges and transverse stiffeners are rigid (see *clause 6.7.8.3*). The web panel is then the panel between two adjacent transverse stiffeners.

Table 6.12 Summary of the shear buckling coefficient

Stiffener	Shear buckling coefficient	
No intermediate stiffeners	$k_\tau = 5.34$	
Rigid TS, any α	$k_\tau = 5.34 + 4.00(b_w/a)^2$ if $a/b_w \geq 1$ $k_\tau = 4.00 + 5.34(b_w/a)^2$ if $a/b_w < 1$	
1 or 2 LS, $\alpha \geq 3$	$k_\tau = 5.34 + 4.00(b_w/a)^2 + k_{\tau st}$ if $a/b_w \geq 1$	(6.127)
	$k_\tau = 4.00 + 5.34(b_w/a)^2 + k_{\tau st}$ if $a/b_w < 1$	(6.128)
>2 LS, any α	where $k_{\tau st} = 9\left(\dfrac{b_w}{a}\right)^2\left(\dfrac{I_{st}}{t_w^3 b_w}\right)^{3/4}$ but not less than $\dfrac{2.1}{t_w}\left(\dfrac{I_{st}}{b_w}\right)^{1/3}$	(6.129)
1 or 2 LS, $\alpha < 3$	$k_\tau = 4.1 + \dfrac{6.3 + 0.18 I_{st}/(t_w^3 b_w)}{a^2} + 2.2\left(\dfrac{I_{st}}{t_w^3 b_w}\right)^{1/3}$	(6.129a)

TS, transverse stiffener; LS, longitudinal stiffener.

Buckling coefficient k_τ

Expressions for the shear buckling coefficient k_τ in *expression 6.126* are summarised in Table 6.12, where:

- a is the distance between transverse stiffeners (see Figure 6.53).
- I_{st} is the second moment of the area of the longitudinal stiffener with regard to the z axis (see Figure 6.58(b)). For webs with two or more equal stiffeners, not necessarily equally spaced, I_{st} is the sum of the stiffness for the individual stiffeners.

$\alpha = a/b_w$.

The second moment of the area I_{st} is determined with an effective plate width $15t_w$ on each side of the stiffener up to existing geometrical width without overlapping parts. I_{st} is determined for buckling perpendicular to the plane of the web. For stiffened panels with two or more longitudinal stiffeners, I_{st} is the sum of all individual stiffeners regardless of whether they have an equidistant spacing or not.

In order to apply the buckling curves according to expressions D6.48 and D6.49 for a stiffened panel, a reduction in the second moment of the area of the longitudinal stiffener I_{st} to one-third of its actual value is required when calculating k_τ. This accounts for the reduced post-critical reserve of stiffened panels compared with unstiffened plates. *Expressions 6.129 and 6.129a* already consider the one-third reduction in the moment of inertia of the longitudinal stiffeners.

For intermediate non-rigid transverse stiffeners, it is stated in *clause 6.7.8.3(2)* that their stiffness should be considered in the calculation of k_τ, but no expressions are given. However, in aluminium structures, intermediate non-rigid transverse stiffeners are rarely used in practice, since the increase in shear resistance may be very low. Even intermediate rigid transverse stiffeners are not advantageous if shear resistance and the reduction in web thickness are traded off against the additional cost of welding. On the other hand, longitudinal stiffeners may be added to the profile when extruded with little extra cost. See, for example, Example 6.3, where a screw port is used as a stiffener without adding any material.

Clause 6.7.8.3(2)

Figure 6.57 shows a web with transverse and longitudinal stiffeners.

Figure 6.57. (a) Web with transverse and longitudinal stiffeners. (b) Effective cross-section of longitudinal stiffeners. (Reproduced from EN 1999-1-1 (*Figure 6.29*), with permission from BSI)

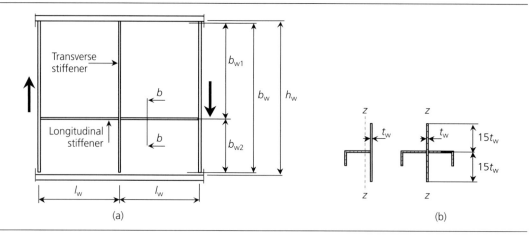

(a)   (b)

Clause 6.7.4.2(10)

Contribution from the flanges (*clause 6.7.4.2(10)*)

The contribution from the flanges is usually negligible for ordinary aluminium plate girders. However, if the distance between the transverse stiffeners is small, the contribution can be accounted for according to *expression 6.131*, which assumes the formation of four plastic hinges in the flanges at a distance c (Figure 6.58):

$$V_{f,Rd} = \frac{b_f t_f^2 f_{of}}{c\gamma_{M1}}\left[1 - \left(\frac{M_{Ed}}{M_{f,Rd}}\right)^2\right] \tag{6.131}$$

in which b_f and t_f are for the flange leading to the lowest resistance, b_f being taken as not larger than $15t_f$ on each side of the web, and $M_{f,Rd}$ is the design moment resistance of the cross-section considering the effective flanges only. Based on tests, evaluated in Höglund (1995),

$$c = \left(0.08 + \frac{4.4 b_f t_f^2 f_{of}}{t_w b_w^2 f_{ow}}\right)a \tag{6.131a}$$

The contribution of the flanges is reduced if they resist longitudinal stresses due to the bending moment M_{Ed} according to the last term in *expression 6.131*. The design bending moment resistance $M_{f,Rd}$ consists of a cross-section with the effective area of the flanges only. *Expression 6.131* corresponds to curve 1 in Figure 6.59.

Influence of the bending moment

If the girder is subjected to a shear force and at the same time to a small bending moment ($M_{Ed} < M_{f,Rd}$, where $M_{f,Rd} = A_f d f_{of}$ is the moment capacity of the flanges and d is the distance between the centre of the flanges), then it is assumed that the stresses in the web that are caused by the bending moment do not influence that portion $V_{w,Rd}$ of the shear force that

Figure 6.58. Model of a web in the post-buckling range. (a) Shear force carried by the web by the rotated stress field. (b) Shear force carried by the flanges. (c) Notation for the cross-section

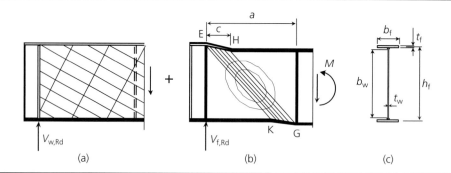

(a)   (b)   (c)

Figure 6.59. Interaction diagram for a girder subjected to the shear force and the bending moment. (Reproduced from EN 1999-1-1 (*Figure 6.32*), with permission from BSI)

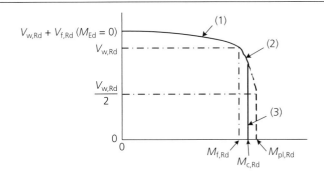

is resisted by the web. On the other hand, if $M_{Ed} > M_{f,Rd}$, then the flanges cannot contribute to the shear capacity of the girder, and the capacity of the web to carry shear forces is reduced. The interaction formula given by Basler (1961) is applied (curve 2 in Figure 6.59):

$$\frac{M_{Ed} + M_{f,Rd}}{2M_{pl,RD}} \frac{V_{Ed}}{V_{w,Rd}}\left(1 - \frac{M_{f,Rd}}{M_{pl,Rd}}\right) \leq 1.00 \tag{6.147}$$

where:

$M_{o,Rd}$ is the design bending moment resistance according to ***clause 6.7.2(4)***
$M_{f,Rd}$ is the design bending moment resistance of the flanges only (=$\min(A_{f1}, A_{f2})h_f f_o/\gamma_{M1}$)
$M_{pl,Rd}$ is the plastic design bending moment resistance.

Clause 6.7.2(4)

Clearly, the bending moment should not exceed the bending moment resistance indicated by the vertical line 3 in Figure 6.59, $M_{Ed} \leq M_{o,Rd}$.

Influence of axial force
$M_{f,Rd}$ in *expressions 6.131* and *6.147* should be reduced with a factor according to *expression 6.132* if a normal force N_{Ed} is also acting:

$$\left(1 - \frac{N_{Ed}}{(A_{f1} + A_{f2})f_o/\gamma_{M1}}\right) \tag{6.132}$$

where A_{f1} and A_{f2} are the areas of the top and bottom flanges, respectively. Thus, *expressions 6.131* and *6.132* consider the interaction between the shear force, the bending moment and the normal force for $M_{Ed} < M_{f,Rd}$. If $M_{Ed} > M_{f,Rd}$, $M_{pl,Rd}$ in *expression 6.147* should be replaced by a reduced plastic moment resistance, given by *expression 6.148*:

$$M_{N,Rd} = M_{pl,Rd}\left[1 - \left(\frac{N_{Ed}}{(A_{f1} + A_{f2})f_o/\gamma_{M1}}\right)^2\right] \tag{6.148}$$

Influence of welds
As there is no web deflection close to the flanges, the membrane stresses in the web close to the flanges are more or less shear stresses only (for a web in shear). In this case, reduced strength in the HAZs due to welding of the web to the flanges does not seem to influence the shear strength of the web very much (Edlund *et al.*, 2001). Therefore, in the above design expressions, the yield strength of the parent material is used and not the strength in the HAZ. Furthermore, it should be noted that all tests supporting the design rules are made on welded girders.

6.7.5   Web stiffeners
Rigid end post (*clause 6.7.8.1*)
The rigid end post act as a bearing stiffener resisting the reaction at the girder support and as a short beam resisting the longitudinal membrane stresses in the plane of the web. It is assumed that the reaction force is acting on the inner stiffeners, as shown in Figure 6.60(a).

Clause 6.7.8.1

Figure 6.60 (a) Rigid end post. (b, c) Non-rigid end posts. (d) Inner support of a continuous girder

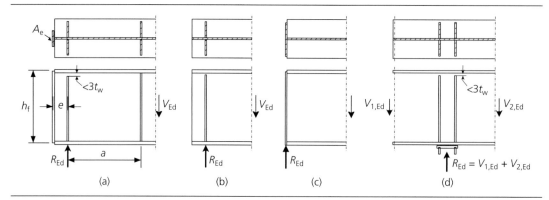

A rigid end post may comprise of one stiffener at the girder end and an inner, double-sided transverse stiffener, together forming the flanges of a short beam of length h_f (see Figure 6.60(a)). The strip of web plate between the stiffeners forms the web of the short beam. Alternatively, an end post may be in the form of an inserted section, connected to the end of the web plate.

The rigid end post should resist the longitudinal tensile stresses. Based on bending of the end post, a requirement for the cross-section area of the extra stiffener is derived (Höglund, 1971). This requirement is simplified to the following rule of thumb:

$$A_e \geq 4h_f t_w^2/e \quad \text{and} \quad e > 0.1h_f \tag{D6.51}$$

where e is the distance between the stiffeners (see Figure 6.60(a)). The inner stiffener is designed for the reaction force R_{Ed} as a load-bearing stiffener.

If an end post is the only means of providing resistance against twist at the end of a girder, the second moment of the area of the end-post section about the centre-line of the web should, according to ***clause 6.7.8.1(6)***, satisfy

Clause 6.7.8.1(6)

$$I_{ep} \geq b_w^3 t_f R_{Ed}/250 W_{Ed} \tag{6.152}$$

where:

t_f is the maximum value of flange thickness along the girder
R_{Ed} is the reaction at the end of the girder under design loading
W_{Ed} is the total design loading on the adjacent span.

A gap, smaller than three times the web thickness, may be left between the top of the stiffener and the upper flange, as shown in Figures 6.60(a) and 6.60(b), unless there is a load on the top flange.

Clause 6.7.8.2

Non-rigid end post (*clause 6.7.8.2*)
A non-rigid end post may be a single pair of stiffeners (as shown in Figure 6.60(b)) or a single stiffener at the girder end (as in Figure 6.60(c)). It is assumed to act as a bearing stiffener resisting the reaction at the girder support. If the end post is the only means of providing resistance against twist at the end of a girder, the requirement in *expression 6.152* should be satisfied.

Clause 6.2.11

Bearing stiffeners at an inner support of a continuous girder (*clause 6.2.11*)
Bearing stiffeners of an inner support can consist of two pairs of stiffeners (as in Figure 6.60(d)) or a single pair of stiffeners. It is assumed to act as a bearing stiffener resisting the reaction at the girder support. In both cases, it is assumed to act as a rigid end-post when calculating the shear strength $V_{1,Rd}$ and $V_{2,Rd}$ for the adjacent web panels.

Also in this case, a gap, smaller than three times the web thickness, can be left between the top of the stiffeners and the upper flange (as shown in Figure 6.60(d)), unless there is a load on the top flange.

Intermediate transverse stiffeners (*clause 6.7.8.3*)
Intermediate stiffeners, acting as **rigid supports of interior panels** of the web, should be checked for resistance and stiffness.

Other intermediate transverse stiffeners may be considered as **flexible stiffeners**, their stiffness being considered in the calculation of k_τ. No expressions are given in EN 1999-1-1 for this case. If values from the literature are used, the stiffness should be reduced to one-third of its actual value (see under 'Buckling coefficient', above).

The stiffeners can consist of a stiffener on one side of the web only, or pairs of stiffeners (see Figure 6.60(b)). Intermediate transverse stiffeners acting as rigid supports for the web panel should have a second moment of area fulfilling the following:

$$I_{st} \geq 1.5 h_w^3 t_w^3 / a^2 \qquad \text{if } a/h_w < \sqrt{2} \tag{6.153}$$

$$I_{st} \geq 0.75 h_w t_w^3 \qquad \text{if } a/h_w \geq \sqrt{2} \tag{6.154}$$

The second moment of the area I_{st} should be calculated for a cross-section consisting of the stiffener itself and strips of the web, width $15 t_w$, on both side of the stiffener. See Figure 6.57(b).

The strength of intermediate rigid stiffeners should, according to ***clause 6.7.8.3(3)***, be checked for
an axial force $N_{st,Ed}$ according to

$$N_{st,Ed} = V_{Ed} - \rho_{v,0} b_w t_w f_{ow} / \gamma_{M1} \tag{D6.52}$$

where $\rho_{v,0}$ is the reduction factor of the web with the considered stiffener removed. This is a compromise of the force corresponding to a pure diagonal tension field theory and the rotated stress field theory, the latter giving a compression force equal to the shear force $V_{f,Rd}$ carried by the flanges.

If the result is negative, there is no special requirement for strength, only the requirements for stiffness given above. In the case of variable shear forces, the check is performed for the shear force at a distance $0.5 h_w$ from the edge of the panel with the largest shear force.

Longitudinal stiffeners (*clause 6.7.8.4*)
Longitudinal stiffeners may be either rigid or flexible. In both cases their stiffness should be taken into account when determining the buckling coefficient k_τ in the expressions in Table 6.12. This means that if a sub-panel governs the value of λ_w, then the stiffener may be considered as rigid.

The strength should be checked for axial stresses if the stiffeners are taken into account for resisting such stresses due to bending or an axial load on the girder.

Welds (*clause 6.7.8.5*)
The web-to-flange welds may be designed for the nominal shear flow V_{Ed}/h_w if V_{Ed} does not exceed $\rho_v h_w t_w f_o / (\sqrt{3} \gamma_{M1})$. For larger values, the weld between flanges and webs should be designed for the shear flow $\eta t_w f_o / (\sqrt{3} \gamma_{M1})$ unless the state of stress is investigated in detail.

6.7.6   Resistance to transverse loads
Transverse loading denotes a load that is applied perpendicular to the web. Sometimes the loading is free and transient, as for crane runway girders, where transverse stiffeners are not appropriate. A concentrated transverse loading is often referred to patch loading.

The collapse behaviour of girders subjected to transverse loading have been characterised by three failure modes: yielding, crippling or buckling. In reality, no separation of these phenomena is possible, so in ***clause 6.7.5*** of EN 1999-1-1 (as in EN 1993-1-5) they are merged into a single
verification based on Lagerqvist (1994) and further verified for aluminium girders by Tryland (1999). Their approach presumes that the load is introduced into the plate via the flanges; thus, it should not be applied to single plates under patch load.

Table 6.13. Load application, buckling coefficient and effective loaded length

Type	Load application	Buckling coefficient	Effective loaded length
(a)	$V_{1,Ed}$ S_s F_{Ed} $V_{2,Ed}$ a $V_{1,Ed} + V_{2,Ed} = F_{Ed}$	$k_F = 6 + 2\left(\dfrac{h_w}{a}\right)^2$	$l_y = l_{y0} = S_s + 2t_f\left(1 + \sqrt{m_1 + m_2}\right)$ (6.143) but $l_y \leq$ distance between adjacent transverse stiffeners
(b)	F_{Ed} S_s F_{Ed}	$k_F = 3.5 + 2\left(\dfrac{h_w}{a}\right)^2$	
(c)	c F_{Ed} S_s V_{Ed} $V_{Ed} = F_{Ed}$	$k_F = 2 + 6\dfrac{S_s + c}{h_w} \leq 6$	$l_y = \min(l_{y0}, l_{y1}, l_{y2})$, where l_{y0} from above $$l_{y1} = l_e + t_f\sqrt{\dfrac{m_1}{2} + \left(\dfrac{l_e}{t_f}\right)^2 + m_2} \quad (6.144)$$ $$l_{y2} = l_e + t_f\sqrt{m_1 + m_2} \quad (6.145)$$ where $$l_e = \dfrac{k_F E t_w^2}{2f_{ow}h_w} \leq S_s + c \quad (6.146)$$

Clause 6.7.5

The rules in *clause 6.7.5* cover both extruded and welded girders, with and without longitudinal stiffeners. It is assumed that the compression flange has an adequate lateral and torsional restraint.

Clause 6.7.5

Clause 6.7.5 covers three different types of transverse load application, as follows:

- forces applied through one flange and resisted by shear forces in the web: '*patch loading*' (Table 6.13, type (a))
- forces applied to one flange and transferred through the web directly to the other flange: '*opposite patch loading*' (Table 6.13, type (b))
- forces applied through one flange adjacent to an unstiffened end: '*end patch loading*' (Table 6.13, type (c)).

For cross-sections with inclined webs, it has to be taken into account that transverse stresses are not only induced in the web but also in the bottom plate, so that the corresponding in-plane components of the transverse loading have to be considered. See Figure 6.61(b).

Figure 6.61. (a) Buckling of the web and plastic hinges in the flanges under a concentrated load. (b) Forces in inclined webs. (c) Length of the stiff bearing and the effective loaded length

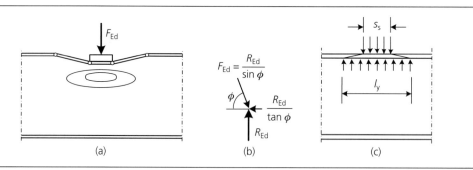

The design approach in ***clause 6.7.5*** follows the same procedure as for other stability problems, including three parameters:

Clause 6.7.5

- the plastic resistance F_o
- the elastic buckling force F_{cr}, which defines a slenderness $\lambda_F = \sqrt{F_o/F_{cr}}$
- the reduction factor $\chi_F = \chi_F(\lambda_F)$, which reduces the yield resistance for λ_F larger than a certain limiting value such that $F_R = \chi_F F_o$.

The simple expression $\chi_F = 0.5/\lambda_F$ (but not more than 1.0) is used for χ_F. This gives conservative results (Lagerqvist, 1994; Tryland, 1999). The simple expression for χ_F means that the resistance can be given in one simple formula:

$$F_R = \chi_F F_o = \frac{0.5}{\lambda_F} F_o = 0.5\sqrt{F_{cr} F_o} \qquad \text{but not more than } F_o \tag{D6.53}$$

where

$$F_o = l_y t_w f_{ow} \tag{D6.54}$$

$$F_{cr} = 0.9 k_F E t_w^3 h_w \tag{6.138}$$

and l_y is the effective loaded length (Figure 6.61(c)), obtained from ***clause 6.7.5.5***.

Clause 6.7.5.5

For webs without longitudinal stiffeners the factor k_F should be obtained from Table 6.13.

For webs with longitudinal stiffeners, k_F should be taken as

$$k_F = 6 + 2(h_w/a)^2 + (5.44 b_1/a - 0.21)\sqrt{\gamma_s} \tag{6.139}$$

where b_1 is the depth of the loaded sub-panel taken as the clear distance between the loaded flange and the stiffener and

$$\gamma_s = 10.9 I_{sl}/(h_w t_w^3) \le 13(a/h_w)^3 + 210(0.3 - b_1/h_w) \tag{6.140}$$

where I_{sl} is the second moment of the area (about the z–z axis) of the stiffener closest to the loaded flange, including contributing parts of the web according to Figure 6.57. *Expression 6.140* is valid for $0.05 \le b_1/h_w \le 0.3$ and loading according to type (a) in Table 6.13.

Inserting expression D6.54 and *expression 6.138* into expression D6.53, and introducing the resistance factor γ_{M1}, the design resistance F_{Rd} for a transverse force is found as

$$F_{Rd} = 0.57 t_w^2 \sqrt{\frac{k_F l_y f_{ow} E}{b_w}} \frac{1}{\gamma_{M1}} \qquad \text{but not more than } t_w l_y \frac{f_{ow}}{\gamma_{M1}} \tag{D6.55}$$

where k_F and l_y are given in Table 6.13.

The length of stiff bearing, s_s, on the flange is the distance over which the applied force is effectively distributed. It may be determined by dispersion of the load through solid material at a slope of 1:1 (see Figure 6.62). s_s should not be taken as more than h_w.

Figure 6.62. Length of stiff bearing. (Reproduced from EN 1999-1-1 (*Figure 6.31*), with permission from BSI)

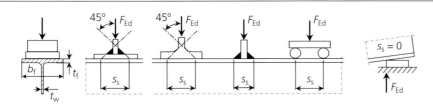

If several concentrated loads are closely spaced, the resistance should be checked for each individual load as well as for the total load. In the latter case, s_s should be taken as the centre distance between the outer loads.

If the loading device does not follow the change in the slope of the girder end, $s_s = 0$.

The effective loaded length l_y as given in Table 6.13 is calculated using the two dimensionless parameters m_1 and m_2 obtained from

$$m_1 = \frac{f_{of} b_f}{f_{ow} t_w} \tag{6.141}$$

$$m_2 = 0.02 \left(\frac{h_w}{t_f}\right)^2 \quad \text{if } \lambda_F < 0.5 \quad \text{otherwise } m_2 = 0 \tag{6.142}$$

where b_f is the flange width (see Figure 6.62). For box girders, b_f in *expression 6.141* is limited to $15 t_f$ on each side of the web. As m_2 is dependent on whether $\lambda_F > 0.5$ or not, estimation and checking is required, and possibly recalculating after λ_F, if calculated, where:

$$\lambda_F = \sqrt{\frac{l_y t_w w_{ow}}{0.9 k_f E t_w^3 / h_w}} \tag{6.137}$$

6.7.7. Flange induced buckling

When a girder is subject to bending, the curvature combined with the compression in the flange leads to a transverse force applied to the web plane according to Figure 6.63. If the web is slender, buckling may occur. This web-buckling phenomenon is modelled as column buckling of a transverse web stripe, and leads to the following requirement for the ratio b_w/t_w (*clause 6.7.7*):

Clause 6.7.7

$$\frac{b_w}{t_w} \leq \frac{kE}{f_{of}} \sqrt{\frac{A_w}{A_{fc}}} \tag{6.150}$$

where A_w is the area of the web and A_{fc} is the area of the compression flange. The curvature is larger if a plastic moment is utilised, and especially if plastic rotation is utilised. Therefore, the value of k should be taken as follows:

- plastic rotation utilised   0.3
- plastic moment resistance utilised   0.4
- elastic moment resistance utilised   0.55.

If the girder is curved in elevation, with the compression flange on the concave face, the curvature is increased, and thus also the transverse compression force. The following expression should then be checked as well:

$$\frac{b_w}{t_w} \leq \frac{kE}{f_{of}} \sqrt{\frac{A_w}{A_{fc}}} \frac{1}{\sqrt{1 + \frac{b_w E}{3 r f_{of}}}} \tag{6.151}$$

Figure 6.63. Flange-induced buckling

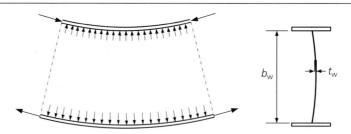

where r is the radius of curvature of the compression flange. If the compression flange is on the convex side, the web will be loaded in tension, and thus there is no risk of flange-induced web buckling. However, in both cases, flange curling for wide flanges needs to be allowed for. See Section 10.8 in this guide and *clause 5.4* of EN 1999-1-4.

Clause 5.4

6.8.   Members with corrugated webs
For plate girders with trapezoidal corrugated webs (Figure 6.64), the bending moment resistance is given in *clause 6.8.1* and the shear force resistance in *clause 6.8.2*.

Clause 6.8.1
Clause 6.8.2

For transverse loads, the rules in *clause 6.7.7* can be used as a conservative estimate.

Clause 6.7.7

Cut outs, even if small, may have a large influence on the shear resistance. No rules are, however, given in the code for corrugated webs with holes or cut-outs.

As for plane webs, many design methods are proposed in the literature (Höglund, 1997). The background to the final choice is given in Johansson *et al.* (2007).

6.8.1   Bending moment resistance
As the web is corrugated, it has no ability to sustain longitudinal stresses, so the contribution is ignored. The bending moment resistance is therefore simply the smallest axial resistance of the flanges times the distance between the centroids of the flanges. This axial resistance may be influenced by lateral torsional buckling if the compression flange is not braced closely enough. In *clause 6.8.1* the reduction factor for lateral torsional buckling is used, as there are some favourable properties compared with flat webs that are disregarded (e.g. substantial transverse stiffness of the corrugated web) and increase in warping stiffness due to the corrugation. It is not defined in EN 1999-1-1 how to obtain the cross-section properties to calculate the elastic lateral torsional buckling load, but it should be safe to assume a section with a centric plane web.

Clause 6.8.1

The corrugated web also influences the local (torsional) buckling of the compression flange. No recommendations are given in the code on how to define the slenderness b/t. A safe estimate is to use the larger outstand, or the average outstand if

$$\frac{(a_1 + a_2)a_3}{(a_1 + 2a_4)b_1} < 0.14$$

(D6.56)

where b_1 is the width of the compression flange and the other notation is as in Figure 6.64.

If there is a substantial shear force in the cross-section of the maximum bending moment, there may be an influence on the axial resistance of the flange due to the shear flow introduction in the

Figure 6.64. Corrugated web. (Reproduced from EN 1999-1-1 (*Figure 6.33*), with permission from BSI)

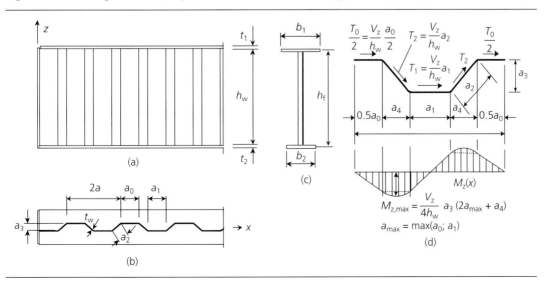

flanges, as indicated in Figure 6.64(d). In *clause 6.8.1* the axial force resistance of the flanges is therefore reduced with a factor

$$\rho_z = 1 - 0.4\sqrt{\frac{\sigma_x(M_z)}{f_{of}/\gamma_{M1}}} \tag{6.156}$$

where M_z is the transverse bending moment in the flange according to Figure 6.64(d) and f_{of} is the 0.2% proof strength of the flange material. Thus, the bending moment resistance may be derived from

$$M_{Rd} = \min\begin{Bmatrix} b_2 t_2 h_f \rho_z f_{of}/\gamma_{M1} \\ b_1 t_1 h_f \rho_z f_{of}/\gamma_{M1} \\ b_1 t_1 h_f \chi_{LT} f_{of}/\gamma_{M1} \end{Bmatrix} \begin{array}{l} \text{tension flange} \\ \text{compression flange} \\ \text{compression flange} \end{array} \tag{6.155}$$

where χ_{LT} is the reduction factor for lateral torsional buckling according to *clause 6.3.2*.

As the web is thin compared with flat webs, and the weld is usually performed on one side of the web only, the HAZ is ignored in the code.

6.8.2 Shear resistance

The shear resistance of plate girders with transverse reinforcement in the form of corrugations or closely spaced transverse stiffeners ($a/b_w < 0.3$) may cause the flat parts between stiffeners to buckle locally and the transverse reinforcement to deform with the web in an overall buckling mode.

The shear force resistance V_{Rd} is taken as

$$V_{Rd} = \rho_c t_w h_w \frac{f_o}{\sqrt{3}\gamma_{M1}} \tag{6.157}$$

where ρ_c is the smallest of the reduction factors for local buckling $\rho_{c,l}$, the reduction factor for global buckling $\rho_{c,g}$ and the HAZ softening factor $\rho_{o,haz}$. There might be an interaction between global and local buckling, but it is weak and disregarded in the code (Johansson *et al.*, 2007).

The reason behind the two buckling checks is that local buckling is expected to show a post-critical strength while global buckling is not. This is reflected by $\rho_{c,l}$ appearing linear and $\rho_{c,g}$ squared in the reduction factor. See Table 6.14 and Figure 6.65.

Table 6.14. Reduction factors for corrugated webs

	Local buckling		Global buckling		HAZ softening
Reduction factor	$\rho_{c,l} = \dfrac{1.15}{0.9 + \lambda_{c,l}} \le 1.0$	(6.158)	$\rho_{c,g} = \dfrac{1.15}{0.5 + \lambda_{c,g}^2} \le 1.0$	(6.160)	$\rho_{o,haz}$ *Clause 6.1.6*
Slenderness	$\lambda_{c,l} = 0.35\dfrac{a_{max}}{t_w}\sqrt{\dfrac{f_o}{E}}$	(6.159)	$\lambda_{c,g} = \sqrt{\dfrac{f_o}{\sqrt{3}\tau_{cr,g}}}$	(6.161)	
Notation	$a_{max} = \max(a_0, a_1, a_2)$		$\tau_{cr,g} = \dfrac{32.4}{t_w h_w^2}\sqrt[4]{B_x B_z^3}$	(6.162)	
	a_0, a_1 and a_2 are widths of folded web panels (see Figure 6.64)		$B_x = \dfrac{2a}{a_0 + a_1 + 2a_2}\dfrac{E t_w^3}{10.9}$		
			$B_z = \dfrac{E I_x}{2a}$		
			I_x is the second moment of the area of one corrugation of length $2a$		

Figure 6.65. Reduction factors for corrugated webs ($\rho_{o,haz}$ is dependent on the alloy, and is given here as an example)

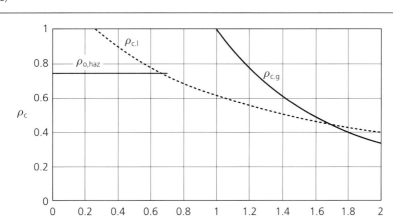

Example 6.17: plate girder in shear, bending and concentrated forces

A simply supported primary plate girder supports four secondary beams, as shown in Figure 6.66. Full lateral restraint is assumed at the load application points B, C, D and E. The loads are $F_{Ed} = 450$ kN. The plate girder is made of two T section extrusions and a plate welded together into an I section.

Figure 6.66. General arrangement, loading, shear and moment diagrams, cross-section and details under the concentrated load and at the girder ends

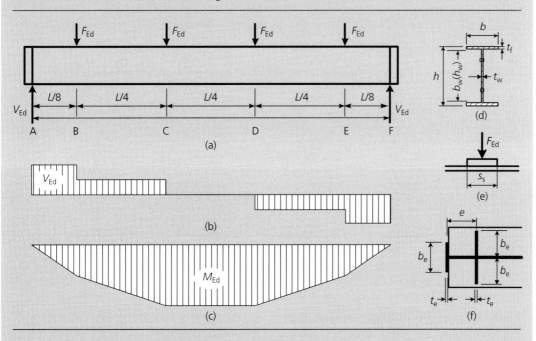

The loading, shear force and bending moment diagrams are shown in Figure 6.66. A lateral torsional buckling check will be carried out on segment CD. Flange-induced buckling, shear and patch loading will be checked, as well as deflections at the serviceability limit state.

Cross-section properties and bending moments

Span	$L = 8.2$ m
Segment length	$c_p = 2.05$ mm
Section height	$h = 1000$ mm

121

Flange width	$b = 320$ mm
Web thickness	$t_w = 12$ mm
Flange thickness	$t_f = 25$ mm
Web-flange fillet radius	$r = 10$ mm
Web height	$b_w = h - 2t_f - 2r = 930$ mm
Welds	$a_w = 4$ mm

Material in flange profile	EN AW-6082 T6 $f_o = 260$ MPa (*Table 3.2a*)
Material in web plate	EN AW-6082 T6 $f_{ow} = 255$ MPa (*Table 3.2b*)
Partial safety factor	$\gamma_{M1} = 1.1$

Bending moment at C and D	$M_{Ed} = 2F_{Ed}1.5c_p - F_{Ed}c_p = 2 \times 450 \times 2.05 = 1845$ kN m
Bending moment at B and E	$M_{B,Ed} = 2F_{Ed}c_p = 2 \times 450 \times 2.05/2 = 922$ kN m

Clause 6.1.4

Cross-section classification, y–y axis bending (*clause 6.1.4*)

$$\varepsilon = \sqrt{250/f_o} = \sqrt{250/260} = 0.981$$

Limits for class 2 and 3 for outstands:

$$\beta_{2,o} = 4.5\varepsilon = 4.41$$

$$\beta_{3,o} = 6\varepsilon = 5.88$$

Clause 6.1.4.4(3)

Limits for classes 2 and 3 for the web, buckling class A, with welds (see *clause 6.1.4.4(3)*):

$$\beta_{2,i} = 13\varepsilon = 12.75$$

$$\beta_{3,i} = 18\varepsilon = 17.7$$

Outstand flanges (*expression 6.1*):

$$\beta_f = (b - t_w - 2r)/2t_f = (320 - 12 - 2 \times 10)/2 \times 25 = 5.76$$

The flange is class 3.

Web − internal part (*expression 6.1*):

$$\beta_w = 0.4b_w/t_w = 0.4 \times 930/12 = 31.0$$

The web is class 4.

In y–y axis bending, the overall cross-section classification is class 4. The resistance is therefore based on the section modulus of the effective cross-section. As the flanges belong to class 3 and the cross-section is symmetric, no iteration is needed to find the neutral axis to calculate the compressed part b_c of the web.

Clause 6.7.7

Flange-induced buckling (*clause 6.7.7*)
Elastic moment resistance is utilised, so $k = 0.55$ in *condition 6.151*:

$$\frac{b_w}{t_w} = \frac{950}{12} = 79.2 < \frac{kE}{f_{of}}\sqrt{\frac{h_w t_w}{b t_f}} = \frac{0.55 \times 70000}{260}\sqrt{\frac{950 \times 12}{320 \times 25}} = 177$$

The condition is met.

Clause 6.2.5
Clause 6.1.6.3

Design resistance for y–y axis bending (*clause 6.2.5*)
The reduction in the web thickness due to local buckling is required (*clause 6.1.6.3*).

From Table 6.5 (*Table 6.3*), buckling class A, with welds, $C_1 = 29$ and $C_2 = 198$:

$$\rho_c = C_1 \frac{\varepsilon}{\beta_w} - C_2 \left(\frac{\varepsilon}{\beta_w}\right)^2 = 29 \frac{0.981}{31.0} - 198 \left(\frac{0.981}{31.0}\right)^2 = 0.719$$

$$t_{w,ef} = \rho_c t_w = 0.719 \times 12 = 8.63 \, \text{mm}$$

HAZ softening (*clause 6.1.6.3*), MIG weld:

Clause 6.1.6.3

$$b_{haz} = 30 \, \text{mm}$$

From Table 3.1 (*Table 3.2b*):

$$\rho_{o,haz} = 0.48$$

Due to added material at the welds (Figure 6.67), the reduction factor for HAZ may be increased to

$$\rho_{haz} = \frac{(2b_{haz} - 2.5t_w)t_w + 2.5t_w \times 2.5t_w}{2b_{haz}} \rho_{o,haz} = 0.997$$

This factor could be further increased, as the weld is not in a section of maximum stress (see *clause 6.1.4.4(4)*). Furthermore, the stiffening effect is omitted. The result is that the web may be assumed flat, omitting HAZ softening.

Clause 6.1.4.4(4)

Second moment of the area of the effective cross-section with respect to the y–y axis:

$$I_{eff,0} = b_f t_f (h - t_f)^2 / 2 + t_w (h - 2t_f)^3 / 12 - (t_w - t_{w,ef})b_c^3 / 3$$

$$= 320 \times 25 \times 975^2 / 2 + 12 \times 950^3 / 12 - (12 - 8.63)(930/2)^3 / 3 = 4.548 \times 10^9 \, \text{mm}^4$$

$$A_{eff} = 2b_f t_f + t_w (h - 2t_f) - (t_w - t_{w,ef})b_c$$

$$= 2 \times 320 \times 25 + 12 \times 950 - (12 - 8.63)465 = 2583 \times 10^4 \, \text{mm}^2$$

Shift of the centre of gravity:

$$z_{gc} = (t_w - t_{w,ef})b_c^2 / 2) / A_{eff} = ((12 - 8.63)465^2 / 2) / 2.583 \times 10^4 = 14.1 \, \text{mm}$$

Figure 6.67. Gross cross-section, effective cross-section and welds

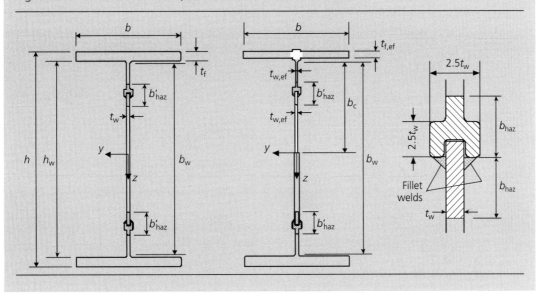

Second moment of the area of the effective cross-section with respect to the centre of gravity of the effective cross-section:

$$I_{\text{eff}} = I_{\text{eff},0} - A_{\text{eff}}z_{\text{gc}}^2 = 4.548 \times 10^9 - 2.583 \times 10^4 \times 14.1^2 = 4.543 \times 10^9 \, \text{mm}^4$$

$$W_{\text{eff}} = I_{\text{eff}}/(h/2 + z_{\text{gc}}) = 4.543 \times 10^9/(500 + 14.1) = 8.836 \times 10^6 \, \text{mm}^3$$

Cross-section resistance for y–y axis bending:

$$M_{\text{y,Rd}} = f_{\text{o}}W_{\text{eff}}/\gamma_{\text{M1}} = 260 \times 8.836 \times 10^6/1.1 = 2089 \, \text{kN m}$$

Clause 6.3.2.1

Lateral torsional buckling of the segment in bending (*clause 6.3.2.1*)

The elastic lateral torsional buckling load is found in *Annex I*. The warping constant and the torsion constant are found in *Annex J*, the torsion constant in *clause J.1* and the warping constant in *Figure J.2* (case 5). Omitting detailed calculations, the constants are

$$I_{\text{z}} = 1.367 \times 10^8 \, \text{mm}^4$$

$$I_{\text{w}} = 3.248 \times 10^{13} \, \text{mm}^6$$

$$I_{\text{t}} = 3.909 \times 10^5 \, \text{mm}^4$$

Clause I.1.1

The elastic critical moment for lateral torsional buckling is, for a constant moment in segment CD with $c_{\text{p}} = L/4 = 2.3$ m, given by the *general formula I.1* in *clause I.1.1* for standard conditions of restraint at each ends, $k_{\text{z}} = 1$, $k_{\text{w}} = 1$ and $k_{\text{y}} = 1$:

$$M_{\text{cr}} = \frac{\pi\sqrt{EI_{\text{z}}GI_{\text{t}}}}{c_{\text{p}}}\sqrt{1 + \frac{\pi^2 EI_{\text{w}}}{c_{\text{p}}^2 GI_{\text{t}}}} = 1.106 \times 10^4 \, \text{kN m} \tag{I.1}$$

The slenderness $\bar{\lambda}_{\text{LT}}$ is

$$\bar{\lambda}_{\text{LT}} = \sqrt{\frac{W_{\text{eff}}f_{\text{o}}}{M_{\text{cr}}}} = \sqrt{\frac{8.836 \times 10^6 \times 260}{11.06 \times 10^9}} = 0.456 \tag{6.58}$$

Clause 6.3.2.2(1)

For class 3 and 4 cross-sections the parameters in the formulae for the reduction factor χ_{LT} for lateral torsional buckling are $\alpha_{\text{LT}} = 0.20$ and $\bar{\lambda}_{0,\text{LT}} = 0.4$ according to *clause 6.3.2.2(1)*, which gives $\chi_{\text{LT}} = 0.986$.

Lateral torsional buckling resistance of segment BC:

$$M_{\text{b,Rd}} = \frac{\chi_{\text{LT}}f_{\text{o}}W_{\text{eff}}}{\gamma_{\text{M1}}} = 2060 \, \text{kN m}$$

$$\frac{M_{\text{Ed}}}{M_{\text{b,Rd}}} = \frac{1845}{2060} = 0.896 < 1.0$$

Segment CD is acceptable.

Clause 6.7.4
Clause 6.7.4.1

Shear force (*clause 6.7.4*)

The design shear resistance is found in *clause 6.7.4.1* for plate girders with web stiffeners at supports only. Rigid end posts are assumed. The strength in the web plate ($f_{\text{ow}} = 255$ MPa) is used.

$$\lambda_{\text{w}} = 0.35\frac{b_{\text{w}}}{t_{\text{w}}}\sqrt{\frac{f_{\text{ow}}}{E}} = 0.35\frac{950}{12}\sqrt{\frac{255}{70\,000}} = 1.64 \tag{6.123}$$

Reduction factor D6.48c (*Table 6.13* in EN 1999-1-1), $\lambda_w \geq 0.937$:

$$\rho_v = \frac{2.3}{1.66 + \lambda_w} = \frac{2.3}{1.66 + 1.65} = 0.694$$

Shear force resistance:

$$V_{Rd} = \rho_v t_w h_w \frac{f_{ow}}{\sqrt{3}\gamma_{M1}} = 0.694 \times 12 \times 950 \frac{255}{\sqrt{3} \times 1.1} = 1064 \, \text{kN} \qquad (6.122)$$

The shear resistance contribution from the flanges according to *clause 6.7.4.2(10)* is neglected as there are no intermediate stiffeners:

Clause 6.7.4.2(10)

$$\frac{V_{Ed}}{V_{Rd}} = \frac{900}{1064} = 0.85 < 1.0$$

The shear resistance is acceptable.

Rigid end post (*clause 6.7.8.1*)

Clause 6.7.8.1

The spacing between the stiffeners should be (see Figure 6.66(f))

$$e > 0.1h_f = 0.1(1000 - 25) = 98 \, \text{mm}$$

say 150 mm.

The area of the stiffener at the end should be

$$A_{st} = 4h_f t_w^2 / e = 4 \times 925 \times 12^2 / 150 = 3550 \, \text{mm}^2$$

for instance 150 × 25 mm.

The second moment of the area of the end-post section about the centre-line of the web should satisfy

$$I_{ep} \geq b_w^3 t_f R_{Ed} / 250 W_{Ed} = 950^3 \times 12 \times 900 / 250 \times 4 \times 450 = 21 \times 10^6 \, \text{mm}^4 \qquad (6.152)$$

Use two 150 × 25 plates on both side of the web:

$$I_{ep} = 25 \times 312^3 / 12 = 63 \times 10^6 \, \text{mm}^4 > 21 \times 10^6 \, \text{mm}^4$$

The condition satisfied.

Interaction between shear force and bending moment (*clause 6.7.6.1*)

Clause 6.7.6.1

At C and D the shear force is $V_{Ed}/2$ (i.e. half of the maximum shear force). Thus, there is no interaction with the bending moment.

At B and E the moment is $M_{Ed}/2$ (i.e. half of the maximum moment). The flange bending moment resistance is

$$M_{f,Rd} = bt_f(h - t_f)f_o / \gamma_{M1} = 320 \times 25 \times 975 \times 260 / 1.1 = 1844 \, \text{kN m}$$

As $M_{Ed}/2 < M_{f,Rd}$, *condition 6.147* need not be checked.

Transverse load (*clause 6.7.5*)

Clause 6.7.5

The length of the stiff bearing $s_s = 100$ mm (see Figure 6.66(e)).

The buckling coefficient in Table 6.13 for $a = L$ is

$$k_F = 6 + 2(h_w/a)^2 = 6 + 2(950/8200)^2 = 6.03$$

The parameters m_1 and m_2 in the expressions for the effective loaded length are

$$m_1 = \frac{f_{of}b_f}{f_{ow}t_w} = \frac{260 \times 320}{255 \times 12} = 27.2 \qquad (6.141)$$

$$m_2 = 0.02\left(\frac{h_w}{t_f}\right)^2 = 0.02\left(\frac{950}{25}\right)^2 = 28.9 \qquad (6.142)$$

Check that the condition $\lambda_F > 0.5$ for m_2 is fulfilled.

The effective loaded length is

$$l_y = s_s + 2t_f(1 + \sqrt{m_1 + m_2}) = 100 + 2 \times 25(1 + \sqrt{27.2 + 28.9}) = 524\,\text{mm} \qquad (6.143)$$

With F_{cr} from *expression 6.138* inserted into *expression 6.13*, we have

$$\lambda_F = \sqrt{\frac{l_y t_w f_{ow}}{0.9 k_f E t_w^3 / h_w}} = \sqrt{\frac{523 \times 12 \times 255}{0.9 \times 6.03 \times 70\,000 \times 12^3 / 900}} = 1.52 \qquad (6.137)$$

which is >0.5. The value of m_2 is okay.

The three *expressions 6.134, 6.135* and *6.136* are merged into

$$F_{Rd} = 0.474 t_w^2 \sqrt{k_f l_y f_{ow} E / h_w} / \gamma_{M1} = 0.474 \times 12^2 \sqrt{6.03 \times 523 \times 255 \times 70\,000 / 950} / 1.1$$

$$= 478\,\text{kN}$$

The resistance must not be larger than the value corresponding to $\chi_F = 1$ in *expression 6.136* inserted in *expressions 6.135* and *6.134*:

$$F_{Rd} = t_w l_y f_{ow} / \gamma_{M1} = 12 \times 523 \times 255 / 1.1 = 1459\,\text{kN}$$

not designing.

$$\frac{F_{Ed}}{F_{Rd}} = \frac{450}{478} = 0.941 < 1.0$$

The resistance to transverse load is acceptable.

Clause 7.2.4

Serviceability limit state (*clause 7.2.4*)
The vertical deflections w_{tot} are defined in EN 1990 by a number of components (see Section 7.2.4 in this guide). In this example, the additional part w_3 of the deflection due to variable load is calculated only.

The variable load at the serviceability limit state is $F_{Ed,ser} = 225$ kN. The deflection is based on the second moment of the area derived with *expression 7.1*, where

$$\sigma_{gr} = \frac{F_{Ed,ser}}{F_{Ed}} \frac{M_{Ed}}{I_{gr}} \frac{h}{2} = \frac{450}{900} \frac{1845 \times 10^6}{4.66 \times 10^9} \frac{1000}{2} = 99\,\text{MPa}$$

$$I_{ser} = I_{gr} - \frac{\sigma_{gr}}{f_o}(I_{gr} - I_{eff}) = 4.66 \times 10^9 - \frac{99}{260}(4.66 \times 10^9 - 4.54 \times 10^9) \qquad (7.1)$$

$$= 4.62 \times 10^9\,\text{mm}^4 \quad I_{ser} = I$$

With the load application points at $a_1 = L/8$ and $a_2 = 3L/8$ from the supports (see Figure 6.66), the deflection will be

$$w_3 = \left[\frac{a_1}{4L}\left(3 - 4\left(\frac{a_1}{L}\right)^2\right) + \frac{a_2}{4L}\left(3 - 4\left(\frac{a_2}{L}\right)^2\right)\right]\frac{F_{Ed,ser}L^3}{6EI}$$

$$= \left[\frac{1}{8 \times 4}\left(3 - 4\left(\frac{1}{8}\right)^2\right) + \frac{3}{8 \times 4}\left(3 - 4\left(\frac{3}{8}\right)^2\right)\right]\frac{225 \times 10^3 \times 8200^3}{6 \times 70\,000 \times 4.62 \times 10^9}$$

$$= 20.5\,\text{mm} = \frac{L}{400}$$

which is less than the limit $L/200$ often used for roof beams (see Table 7.3).

REFERENCES

Basler K (1961) Strength of plate girders under combined bending and shear. *Journal of the Structural Division, ASCE* **87(ST7)**: 181–197.

Bergman SGA (1948) Behaviour of buckled rectangular plates under the action of shearing forces. Thesis, Royal Institute of Technology, Stockholm.

BSI (1991) BS 8118: Structural use of aluminium. Part 1, Code of practice for design. BSI, Milton Keynes.

Edlund S, Jansson R and Höglund T (2001) Shear buckling of Welded Aluminium Girders. *9th Nordic Steel Construction Conference*, Helsinki.

Höglund T (1968) *Approximativ metod för dimensionering av böjd och tryckt stång*. KTH, Stockholm [in Swedish].

Höglund T (1971) *Simply Supported Long Thin Plate I-girders Without Web Stiffeners Subject to Distributed Transverse Load*. IABSE, London.

Höglund T (1973) *Design of Thin Plate I Girders in Shear and Bending with Special Reference to Web Buckling*. Department of Building Statics and Structural Engineering, Royal Institute of Technology, Stockholm. Bulletin 94.

Höglund T (1995) *Strength of Steel and Aluminium Plate Girders – Shear Buckling and Overall Web Buckling of Plane and Trapezoidal Webs. Comparing with Tests*. Department of Structural Engineering, Royal Institute of Technology, Stockholm. Technical Report 1995:4.

Höglund T (1997) Shear buckling resistance of steel and aluminium plate girders. *Thin-walled Structures* **29(1–4)**: 377–380.

Johansson B, Maquoi R, Sedlacek G, Müller C and Beg D (2007) *Commentary and Worked Examples to EN 1993-1-5 'Plated structural elements'*. European Commission, European Joint Research Centre. JRC Scientific and Technical Report.

Lagerqvist O (1994) Patch loading. Resistance of steel girders subjected to concentrated forces. PhD thesis, Division of Steel Structures, Luleå University of Technology, Luleå.

Mazzolani FM (1995) *Aluminium Alloy Structures*, 2nd edn. Spon, London.

Mazzolani FM (ed.) (2003) *Aluminium Structural Design. CISM Courses and Lectures No. 443*. Springer, Wien.

Mazzolani FM, De Matteis G and Mandara A (1996) Classification system for aluminium alloy connections. *Proceedings of the IABSE Colloquium*, Istanbul.

Nethercot DA and Lawson RM (1992) *Lateral Stability of Steel Beams and Columns – Common Cases of Restraint*. Steel Construction Institute, Ascot. P093.

Timoshenko GP and Gere JM (1961) *Theory of Elastic Stability*, McGraw-Hill, New York.

Tryland T (1999) Aluminium and steel beams under concentrated loading. DrIng thesis, Norwegian University of Science and Technology, Trondheim.

Designers' Guide to Eurocode 9: Design of Aluminium Structures
ISBN 978-0-7277-5737-1

ICE Publishing: All rights reserved
http://dx.doi.org/10.1680/das.57371.129

Chapter 7
Serviceability limit states

This chapter discusses serviceability limit states. The material in this chapter is covered in *Section 7* of EN 1999-1-1, and the following clauses are addressed:

- General   *Clause 7.1*
- Serviceability limit states for buildings   *Clause 7.2*

Overall, the coverage of serviceability considerations in EN 1999-1-1 is very limited, with little explicit guidance provided. However, as detailed below, for further information reference should be made to EN 1990, on the basis that many serviceability criteria are independent of the structural material. Clauses 3.4, 6.5 and A1.4 of EN 1990 contain guidance relevant to serviceability; clause A1.4 of EN 1990 (as with the remainder of Annex A1 of EN 1990) is specific to buildings.

The low value of the Young modulus E has a significant influence on the deformations of an aluminium structure, which means that the deformation at the serviceability limit state is often governing. See also Section 7.2.1. The designer should not forget to check the deformation before the final check of the strength at the ultimate limit state.

7.1.   General
Serviceability limit states are defined in clause 3.4 of EN 1990 as those that concern:

- the functionality of the structure or structural members under normal use
- the comfort of the people
- the appearance of the structure.

For buildings, the primary concerns are horizontal and vertical deflections and vibrations.

According to clause 3.4 of EN 1990, a distinction should be made between reversible and irreversible serviceability limit states. Reversible serviceability limit states are those that would be infringed on a non-permanent basis, such as excessive vibration or high elastic deflections under temporary (variable) loading. Irreversible serviceability limit states are those that would remain infringed even when the cause of infringement was removed (e.g. permanent local damage or deformations).

Further, three categories of combinations of loads (actions) are specified in EN 1990 for serviceability checks: characteristic, frequent and quasi-permanent. These are given by Equations 6.14 to 6.16 of EN 1990, and summarised in Table 7.1 (Table AJ.4 of EN 1990), where each combination contains a permanent action component (favourable or unfavourable), a leading variable component and other variable components. Where a permanent action is unfavourable, which is generally the case, the upper characteristic value of a permanent action $G_{kj,\mathrm{sup}}$ should be used; where an action is favourable (such as a permanent action reducing uplift due to wind loading), the lower characteristic value of a permanent action $G_{kj,\mathrm{inf}}$ should be used.

Unless otherwise stated, for all combinations of actions in a serviceability limit state the partial factors should be taken as unity.

- The characteristic combination of actions would generally be used when considering the function of the structure and damage to structural and non-structural elements.

Table 7.1. Design values of actions for use in the combination of actions (data from EN 1990, Table A1.4)

Combination	Permanent action G_d		Variable actions Q_d	
	Unfavourable	Favourable	Leading	Others
Characteristic	$G_{kj,sup}$	$G_{kj,inf}$	$Q_{k,1}$	$\psi_{0,i}Q_{k,i}$
Frequent			$\psi_{1,1}Q_{k,1}$	$\psi_{2,i}Q_{k,i}$
Quasi-frequent			$\psi_{2,1}Q_{k,1}$	$\psi_{2,i}Q_{k,i}$

- The frequent combination would be applied when considering the comfort of the user, the functioning of machinery and avoiding the possibility of ponding of water.
- The quasi-permanent combination would be used when considering the appearance of the structure and long-term effects (e.g. creep).

The purpose of the ψ factors (ψ_0, ψ_1 and ψ_2) that appear in the load combinations of Table 7.2 is to modify characteristic values of variable actions to give representative values for different situations. Numerical values of the ψ factors are given in Table 7.2.

7.2. Serviceability limit states for buildings

It is emphasised in both EN 1999-1-1 and EN 1090 that serviceability limits (e.g. for deflections and vibrations) should be specified for each project and agreed with the client. Numerical values for these limits are not provided in either document, but may be given in the national documents.

7.2.1 Vertical deflections

The vertical deflections w_{tot} are defined in EN 1990 by a number of components, w_c, w_1, w_2 and w_3 as shown in Figure 7.1 (Figure A1.1 of EN 1990), where:

w_c is the precamber in the unloaded structural member
w_1 is the initial part of the deflection under permanent loads
w_2 is the long-term part of the deflection under permanent loads
w_3 is the additional part of the deflection due to variable loads

Table 7.2. Recommended values of ψ factors for buildings (data from EN 1990, Table A1.1)

Action	ψ_0	ψ_1	ψ_2
Imposed loads in buildings, category (see EN 1991-1-1):			
Category A: domestic, residential areas	0.7	0.5	0.3
Category B: office areas	0.7	0.5	0.3
Category C: congregation areas	0.7	0.7	0.6
Category D: shopping areas	0.7	0.7	0.6
Category E: storage areas	1.0	0.9	0.8
Category F: traffic area, vehicle weight $\leq$ 30kN	0.7	0.7	0.6
Category G: traffic area, 30kN $<$ vehicle weight $\leq$ 160kN	0.7	0.5	0.3
Category H: roofs	0	0	0
Snow loads on buildings (see EN 1991-1-3):*			
Finland, Iceland, Norway, Sweden	0.7	0.5	0.2
Remainder of CEN Member States, for sites located at altitude $H > 1000$ m a.s.l.	0.7	0.5	0.2
Remainder of CEN Member States, for sites Located at altitude $H \leq 1000$ m a.s.l.	0.5	0.2	0
Wind loads on buildings (see EN 1991-1-4)	0.6	0.2	0
Temperature (non-fire) in buildings (see EN 1991-1-5)	0.6	0.5	0

Note: the ψ values may be set by the National Annex.
* For countries not mentioned below, see the relevant local conditions.

Figure 7.1. Definitions of vertical deflection. (Reproduced from EN 1990 (Figure A1.1), with permission from BSI)

Table 7.3. Vertical deflection limits

Design situation	Deflection limit
Cantilevers	Length/180
Beams carrying plaster or other brittle finish	Span/360
Other beams (except purlins and sheeting rails)	Span/200
Purlins and sheeting rail	To suit cladding, typically span/200 under permanent load and span/100 under the worst combinations of variable loads

w_{tot} is the total deflection $w_1 + w_2 + w_3$

w_{max} is the remaining total deflection taking into account the precamber $= w_{tot} - w_c$.

In the absence of prescribed limits, those in Table 7.3 may be used for serviceability verifications based on the characteristic combination of actions. In general, the deflection limits should be checked against the total deflection w_{tot}.

The National Annex may define similar limits to those given in Table 7.3, and may propose that permanent actions be taken as zero in serviceability checks. In this case, w_1 and w_2 would be zero, so w_{tot} would be w_3. Example 7.1 illustrates the calculation of vertical deflection of a beam.

As already mentioned, the low value of the Young modulus has a significant influence on the deformations of an aluminium structure. A well-known example is the bending of beams, where the stiffness EI is the governing factor, and $I_{al} = 3I_{steel}$ to arrive at the same stiffness as a steel beam, which is illustrated in Table 7.4. Remember that to arrive at the same stiffness for an aluminium beam with *same height* as a steel beam you must compensate the low value of the elastic modulus with a corresponding increase in the cross-section area resulting in about the same weight.

The above indicates that, in designing aluminium structures, it is often not the strength but, in many cases, the deformation that is the governing factor. So, in building and civil engineering it is frequently the alloy that does not have the highest strength that should be considered. Do not forget to check the deformations at the serviceability limit state before the final check of the strength at the ultimate limit state.

Table 7.4. Weight of beams with same stiffness EI

Profile	I	I	I
Material	Steel IPE 240	Aluminium Same height	Aluminium Increased height
EI: $\times 10^{12}$ N/mm^2	8.17	8.17	8.17
h: mm	240	240	300
Weight: kg/m	30.7	30.3	18.4

Example 7.1: vertical deflection of a beam

A simply supported glass roof beam of span 4.2 m (Figure 7.2) is subjected to the following unfactored loads:

Dead load	8.6 kN/m
Imposed roof load	10.5 kN/m
Snow load	6.8 kN/m

Design an I beam such that the vertical deflection limits of Table 7.3 are not exceeded.

Clause 3.2.5

From *clause 3.2.5* we find $E = 70\,000$ MPa (N/mm^2).

Using the characteristic combination of action of Table 7.2, where the permanent action is unfavourable and an imposed roof load is the leading variable action, we have the serviceability loading

$$E_d = G_k \text{ '+' } Q_{k,1} \text{ '+' } \psi_{0,2} Q_{k,2}$$

From Table 7.1, for snow loads at altitude > 1000 m, $\psi_0 = 0.7$. Therefore,

$$q = 8.6 + 10.5 + 0.7 \times 6.8 = 23.9 \text{ kN/m}$$

Under a uniformly distributed load, the maximum deflection w of a simply supported beam is given by

$$w = \frac{5}{384} \frac{qL^4}{EI}$$

from which the required second moment of area is solved:

$$I_{\text{req}} = \frac{5}{384} \frac{qL^4}{Ew}$$

For a deflection limit of span/360 for a brittle finish (Table 7.3), we get

$$I_{\text{req}} = \frac{5}{384} \frac{qL^4}{Ew} = \frac{5}{384} \frac{23.9 \times 4200^4}{70\,000(4200/360)} = 1.18 \times 10^8 \text{ mm}^4$$

There are no standard I beams in aluminium, so start with choosing the flange slenderness such that there is no reduction due to local buckling. For EN AW-6063 T6, $f_o = 160$ MPa, class A, without weld, we have the slenderness limit $\beta_3 = 6\varepsilon = 7.5$ for $\varepsilon = \sqrt{250/160} = 1.25$.

Choose the flange slenderness

$$\frac{b}{t} = 2\beta_3 = 15.0$$

Figure 7.2. Simply supported beam and beam cross-section

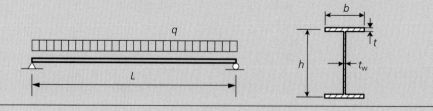

An approximate formula for the second moment of area is

$$I \approx 0.58 A_f h^2$$

where $A_f = bt$, from which, for a chosen value of the beam depth $h = 308$ mm,

$$A_f = \frac{I_{req}}{0.58h^2} = \frac{1.18 \times 10^8}{0.58 \times 308^2} = 2150 \, \text{mm}^2$$

As $b = 2\beta_3 t = 15.0t$, we get

$$t = \sqrt{\frac{A_f}{2\beta_3}} = \sqrt{\frac{5150}{15}} = 12.0 \, \text{mm}$$

$$b = 2\beta_3 t = 15 \times 12.0 = 180 \, \text{mm}$$

Choosing a web thickness $t_w = 6$ mm and checking the resulting second moment of area:

$$I = 2bt\left(\frac{h}{2}\right)^2 + \frac{t_w h^3}{12} = 2 \times 180 \times 12 \times 154^2 + \frac{6 \times 308^3}{12} = 1.17 \times 10^8 \, \text{mm}^4 \approx I_{req}$$

which is acceptable.

As h is the distance between the centres of the flanges, the total height will be

$$308 + 12 = 320 \, \text{mm}$$

7.2.2 Horizontal deflections

Horizontal deflections in structures may be checked using the same combinations of actions as for vertical deflections. The EN 1990 notation to describe horizontal deflections is illustrated in Figure 7.3, where u is the total horizontal deflection of a structure of height H, and u_i is the horizontal deflection in each storey (i) of height H_i.

In the absence of prescribed deflection limits, those provided in Table 7.5 may be used for serviceability verifications based on the characteristic combination of actions.

7.2.3 Dynamic effects

Dynamic effects need to be considered in structures to ensure that vibrations do not impair the comfort of the user or the functioning of the structure or structural members. Essentially, this is achieved provided the natural frequencies of vibration are kept above appropriate levels, which depend upon the function of the structure and the source of vibration. Possible sources of vibration include walking, synchronised movements of people, ground-borne vibrations

Figure 7.3. Definitions of horizontal deflections

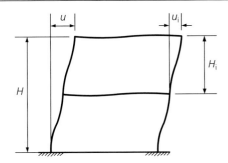

Table 7.5. Horizontal deflection limits

Design situation	Deflection limit
Tops of columns in single-storey buildings, except portal frames	Height/300
Columns in portal frame buildings, not supporting crane runways	To suit cladding
In each storey of a building with more than one storey	Height of storey/300
Curtain wall mullions and transoms for single glazing	Span/175
Curtain wall mullions and transoms for double glazing	Span/250

from traffic, and wind action. Further guidance on dynamic effects may be found in EN 1990 and other specialised literature (Wyatt, 1989).

7.2.4 Calculation of elastic deformation

The calculation of elastic deflection should generally be based on the properties of the gross cross-section of the member. However, for slender sections it may be necessary to take reduced section properties to allow for local buckling (see *clause 6.7.5*). Due allowance for the effects of partitioning and other stiffening effects, second-order effects and changes in geometry should also be made.

Clause 6.7.5

For class 4 sections, the effective second moment of the area I_{ser} may, according to *clause 7.2.4*, be calculated by an interpolation between the second moment of the area of the gross cross-section I_{gr} and the second moment of the area of the effective cross-section at the ultimate limit state I_{eff}, with allowance for local buckling (see *clause 6.2.5.2*):

Clause 7.2.4

Clause 6.2.5.2

$$I_{ser} = I_{gr} - \frac{\sigma_{gr}}{f_o}(I_{gr} - I_{eff}) \qquad (7.1)$$

where σ_{gr} is the maximum compressive bending stress at the serviceability limit state, based on the gross cross-section (positive in the formula).

According to *clause 7.2.4(2)*, the second moment of the area at the serviceability limit state I_{ser} should be taken as constant along the beam, although, in reality, it follows the stress intensity.

Clause 7.2.4(2)

Local weakening due to bolt holes and heat-affected zones may be ignored when calculating deflections. However, due allowance should be made for the rotational stiffness of any semi-rigid joints, and the possible recurrence of local plastic deformation at the serviceability limit state.

Calculation of the deformation of a class 4 cross-section plate girder is included in Example 6.17.

REFERENCE

Wyatt TA (1989) *Design Guide on the Vibrations of Floors*. Steel Construction Institute, Ascot. P076.

Designers' Guide to Eurocode 9: Design of Aluminium Structures
ISBN 978-0-7277-5737-1

ICE Publishing: All rights reserved
http://dx.doi.org/10.1680/das.57371.135

Chapter 8
Design of joints

This chapter discusses joint design, which in *Section 8* of EN 1999-1-1 covers bolts, solid rivets, pins, welds and adhesive bonded joints.

Section 8 of EN 1999-1-1 mirrors EN 1993-1-8 in terms of fundamental material on strengths of bolted and welded joints and information on geometrical constraints. Other information given in EN 1993-1-8 such as joint classification and rules for joints between I sections is mirrored in Annexes B and L.

In many aluminium structures, other joining methods are used as well. Therefore, in this chapter, self-drilling and tapping screws and rivets with a break pull mandrel (blind rivets) from *Section 8* of EN 1999-1-4 ('*Cold-formed structural sheeting*') are included. Some information about screw ports and bolt tracks is given in Section 8.5.16. Other welding methods such as friction stir welding, laser welding and some mechanical fastening methods such as screws in screw port (screw–groove) and self-piercing rivets are mentioned in EN 1999-1-1, but resistance values are not given. It is hoped that design methods for these will be included in subsequent revisions of the code.

Background information to this chapter is given by Soetens in Mazzolani (2002). See also Bulson (1992).

8.1. Basis of design
8.1.1 Introduction
In *clause 8.1.1*, the partial safety factors γ_M for various components in joints are listed. Recommended values are given in Table 8.1. Numerical values for γ_M may be defined in the National Annex.

In *clause 8.1*, general rules and items to pay attention to are listed:

- Applied forces and moments (*clause 8.1.2*). In particular, second-order effects, imperfections, and connection flexibility have to be taken into account.
- Resistance of joints (*clause 8.1.3*).
- Design assumptions (*clause 8.1.4*) allow the internal distribution of forces and moments within the connection in any rational way that gives equilibrium. However, there are limitations on the internal distribution given for bolted connections in *clause 8.5.4* and for welded connections in *clause 8.6.3.5*.
- Fabrication and execution (*clause 8.1.5*).

Clause 8.1.1

Clause 8.1

Clause 8.1.2

Clause 8.1.3
Clause 8.1.4

Clause 8.5.4
Clause 8.6.3.5
Clause 8.1.5

It is pointed out that ease of fabrication and execution should be considered in the design of all joints and splices. Attention should be paid to:

- the clearances necessary for safe execution
- the clearances needed for tightening fasteners
- the need for access for welding
- the requirements of welding procedures
- the effects of angular and length tolerances on fit-up
- the requirements for subsequent inspection
- the requirements for surface treatment
- the requirements for maintenance.

Table 8.1. Recommended partial factors γ_M for joints

Connection	Partial factor
Bolt, rivet and plate connections	$\gamma_{M2} = 1.25$
Pin connections: ■ at the ultimate limit state ■ at the serviceability limit state	$\gamma_{Mp} = 1.25$ $\gamma_{Mp,ser} = 1.0$
Slip resistance: ■ at the ultimate limit state ■ at the serviceability limit state	$\gamma_{Ms,ult} = 1.25$ $\gamma_{Ms,ser} = 1.1$
Welded connections	$\gamma_{Mw} = 1.25$
Adhesive-bonded connections	$\gamma_{Ma} \geq 3.0$
Self-tapping screws and blind rivets (EN 1999-1-4)	$\gamma_{M3} = 1.25$

Requirements for the execution of aluminium structures are given in EN 1090-3.

Either linear elastic or elastic–plastic analysis may be used to determine the forces in the component parts of a joint.

8.2. Intersections for bolted, riveted and welded joints

Clause 8.2

According to *clause 8.2*, members meeting at a joint should usually be arranged with their centroidal axes intersecting at a point. Any kind of eccentricity in the nodes should be taken into account, except in the case of particular types of structures where it has been demonstrated that it is not necessary. The extensive research that has resulted in rules for eccentricity in joints in hollow sections in steel has not been replicated in published format for aluminium members, and it is the authors' view that it would not be wise to use research based on steel joints.

8.3. Joints loaded in shear subject to impact, vibration and/or load reversal

Clause 8.3

Clause 8.5.3

Clause 8.3 gives means to avoid the risk of movement and the loosening of fasteners due to frequent impact or significant vibration or reversal of shear load. For wind and/or stability bracings, bolts in bearing-type connections (category A in *clause 8.5.3*) are allowed.

8.4. Classification of joints

Recommendations for the classification of joints are given in *Annex L*. The explicit links between connection type and methods of global analysis will not be familiar to some designers, and mirrors that given in Eurocode 3. See Section 9.11 in this guide.

8.5. Connections made with bolts, rivets and pins
8.5.1 Positioning of holes for bolts and rivets

Clause 8.5.1

The positioning of holes for bolts and rivets such as to reduce corrosion and local buckling and to facilitate the installation of the bolts or rivets are given in *clause 8.5.1*. Note that the spacing rules do not fully avoid local buckling in compression members, and therefore checks are still necessary. Minimum, regular and maximum spacing and end and edge distances are given in Table 8.2. The minimum values of e_1 and e_2 should be specified with no minus deviation but only plus deviations. Further rules regarding tolerances are given in EN 1090-3: of particular note is the maximum clearance on the diameter of fasteners in normal clearance and oversize holes, which differ from steel and from previous national codes.

For **staggered rows** of fasteners, a minimum line spacing of $p_2 = 1.2d_0$ may be used, provided that the minimum distance s_1 between any two rows of fasteners is greater or equal to $24d_0$ (Figure 8.1(b)).

Maximum values for spacing and edge and end distances are unlimited, except in the following cases (see Table 8.2):

Table 8.2. Minimum and regular end and edge distances and spacing (data from EN 1999-1-1, *Table 8.2*)

Distance or spacing	Minimum	Regular	Maximum for aluminium exposed to the weather or other corrosive influences	Maximum for aluminium not exposed to the weather or other corrosive influences
End distance e_1	$1.2d_0$	$2.0d_0$	$4t + 40$ mm	The larger of $12t$ or 150 mm
Edge distance e_2	$1.2d_0$	$1.5d_0$	$4t + 40$ mm	The larger of $12t$ or 150 mm
Compression members, spacing p_1	$2.2d_0$	$2.5d_0$	The smaller of $14t$ or 200 mm	The smaller of $14t$ or 200 mm
Tension members, spacing p_1	$2.2d_0$	$2.5d_0$	Outer lines: the smaller of $14t$ or 200 mm. Inner lines: the smaller of $28t$ or 400 mm	Outer lines: the smaller of $21t$ or 300 mm. Inner lines: the smaller of $42t$ or 600 mm
Spacing p_2	$2.4d_0$	$3.0d_0$	The smaller of $14t$ or 200 mm	The smaller of $14t$ or 200 mm

- for compression members in order to reduce local buckling and to reduce corrosion in exposed members
- for exposed tension members to reduce corrosion.

The local buckling resistance of the plate in compression between the fasteners should be calculated according to *clause 6.3.1* using $0.6p_1$ as the buckling length. Local buckling between the fasteners need not be checked if $p_1/t \leq 9\varepsilon$. The edge distance should not exceed the local buckling requirements for an outstand element in the compression members (see Sections 6.1.2–6.1.5 in this guide). The end distance is not affected by this requirement.

Clause 6.3.1

Slotted holes are generally not recommended. However, slotted holes may be used for connections in category A with loads perpendicular only to the direction of the slotted hole. Rules are given in *clause 8.5.1(5)*.

Clause 8.5.1(5)

Conditions for **oversized holes** (refer to EN 1090-3 for limits on size) in bolted connections of category A are given in *clause 8.5.1(11)*, and conditions for **countersunk bolts and rivets made of steel** are given in *clause 8.5.7*. No rules are given for countersunk bolts or rivets made of aluminium.

Clause 8.5.1(11)
Clause 8.5.7

8.5.2 Deductions for fastener holes
Design for block-tearing resistance
Figure 8.2 shows several cases of block tearing, which consists of failure in shear at the row of bolts along the shear face of the hole group accompanied by tensile failure along the line of bolt holes on the tension face of the bolt group.

Expressions to cover concentrically and eccentrically loading are provided by *expressions 8.1* and *8.2*, respectively:

$$V_{\text{eff,1,Rd}} = \frac{f_u A_{\text{nt}}}{\gamma_{\text{M2}}} + \frac{f_o A_{\text{nv}}}{\sqrt{3}\gamma_{\text{M1}}} \qquad (8.1)$$

Figure 8.1. (a) Symbols for the positioning of fasteners. (b) Staggered spacing in a joint

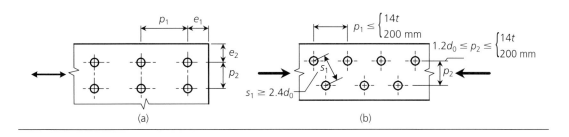

(a)   (b)

Figure 8.2. Block tearing. (Reproduced from EN 1999-1-1 (*Figure 8.5*), with permission from BSI)

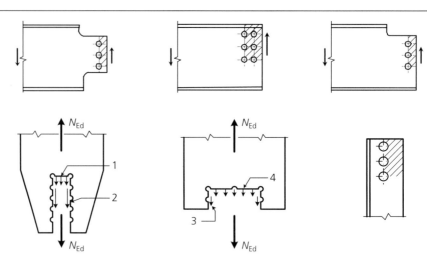

1, small tension force; 2, large shear force; 3, small shear force; 4, large tension force

$$V_{\text{eff,2,Rd}} = \frac{0.5 f_u A_{\text{nt}}}{\gamma_{\text{M2}}} + \frac{f_o A_{\text{nv}}}{\sqrt{3}\gamma_{\text{M1}}} \qquad (8.2)$$

where:

A_{nt} is the net area subjected to tension
A_{nv} is the net area subjected to shear.

Note that the ultimate strength is used for the area in tension but the 0.2% proof strength for the area in shear.

Angles and angles with bulbs

Clause 8.5.2.3

Rules are provided in *clause 8.5.2.3* for the tensile and compressive resistance of angles or angles with bulbs connected through one leg that adopt the usual practice of treating it as concentrically loaded but with a correction factor applied to the area.

8.5.3   Categories of bolted connections

Clause 8.5.3.1

Three categories of shear connections are defined in *clause 8.5.3.1*:

- **Category A: bearing type**, where steel bolts (ordinary or the high-strength type) or stainless steel bolts or aluminium bolts or aluminium rivets should be used. No preloading or special provisions for contact surfaces are required.
- **Category B: slip-resistant at the serviceability limit state**, where preloaded high-strength steel bolts with controlled tightening should be used.
- **Category C: slip resistant at the ultimate limit state**, where preloaded high-strength steel bolts with controlled tightening should be used.

For shear connections where slip resistance is required (categories B and C), treatment of the contact surface is necessary, see Section 8.5.9 in this guide.

Clause 8.5.3.2

Similarly, two categories of bolts used in tension are defined in *clause 8.5.3.2*:

- **Category D: connections with non-preloaded bolts** of steel from class 4.6 up to and including class 10.9 or aluminium or stainless steel. This category should not be used where the connections are frequently subjected to variations in tensile loading. However, they may be used in connections designed to resist normal wind loads.
- **Category E: connections with preloaded high-strength bolts** with controlled tightening. Such preloading improves fatigue resistance. However, the extent of the improvement depends on detailing and tolerances. See EN 1999-1-3, 'Structures susceptible to fatigue'.

Figure 8.3. Example of the distribution of loads between fasteners (five bolts). (a) Elastic load distribution: proportional to the distance from the centre of rotation. (b) Plastic load distribution: possible plastic distribution with one fastener resisting V_{Ed} and four resisting M_{Ed}. (Reproduced from EN 1999-1-1 (*Figure 8.7*), with permission from BSI)

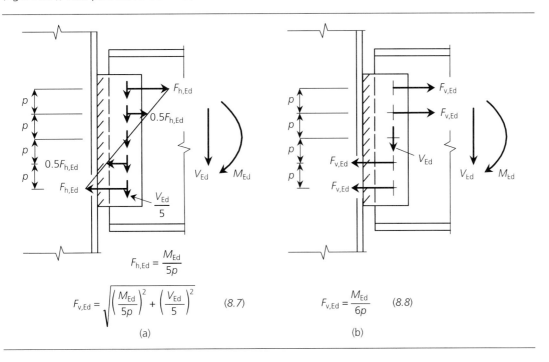

$$F_{h,Ed} = \frac{M_{Ed}}{5p}$$

$$F_{v,Ed} = \sqrt{\left(\frac{M_{Ed}}{5p}\right)^2 + \left(\frac{V_{Ed}}{5}\right)^2} \quad (8.7) \qquad F_{v,Ed} = \frac{M_{Ed}}{6p} \quad (8.8)$$

(a)   (b)

For tension connections in both categories D and E, no special treatment of contact surfaces is necessary, except where connections in category E are subject to both tension and shear (combination E–B or E–C).

Table 8.4 of EN 1999-1-1 lists the design checks needed for each of these five categories.

8.5.4   Distribution of forces between fasteners
Elastic distribution
Clause 8.5.4 requires an elastic distribution in the following two cases:

- category C slip resistant connections
- category A or B connections in cases where the shear resistance $F_{v,Rd}$ of the fastener is less than the design bearing resistance $F_{b,Rd}$, as may occur when bolts pass through very thick elements.

In such cases, the distribution of internal forces between fasteners due to the bending moment at the ultimate limit state should be proportional to the distance from the centre of rotation, and the distribution of the shear force should be equal (Figure 8.3(a)). Such an elastic distribution may also be used for other cases (conservative).

Plastic distribution
In category A and B connections the distribution of internal forces between fasteners due to the bending moment at the ultimate limit state may be assumed plastic, and the shear force may be carried by fasteners not utilised in bending (Figure 8.3(b)).

Factors affecting the resistance of bolted and riveted connections.
The design resistance of connections is reduced in some cases, and specific provisions are stipulated for the following:

- prying action – see *clause 8.5.10*
- long joints – see *clause 8.5.11*
- single lap joints – see *clause 8.5.12*
- fasteners through packing – see *clause 8.5.13*.

Clause 8.5.4

Clause 8.5.10
Clause 8.5.11
Clause 8.5.12
Clause 8.5.13

8.5.5   Design resistances of bolts

The shear, bearing and tension resistances of non-preloaded bolts of steel, stainless steel and aluminium are given in *Table 8.5*, together with requirements for punching shear and combined shear and tension. In general, the bolts and corresponding nuts and washers should be in accordance with the materials and standards listed in *Table 3.4* and, to the reference standards listed in EN 1090-3. Provision is also given for calculating the resistance of items such as holding down bolts or tie rods that have had threads cut in standard round bars by applying a reduction factor to the values in *Table 8.5* (*clause 8.5.5(6)*) As there is no European standard for aluminium bolts, the additional requirements of *clause C.4.1* should be followed.

Clause 8.5.5(6)
Clause C.4.1

Shear resistance per shear plane

$$F_{v,Rd} = \frac{\alpha_v f_{ub} A}{\gamma_{M2}} \qquad (8.9)$$

where:

$\alpha_v = 0.6$ for steel bolts with classes 4.6, 5.6 and 8.8 if the shear plane passes through the threaded portion of the bolt, and for all classes if the shear plane passes through the unthreaded portion of the bolt

$\alpha_v = 0.5$ for steel bolts with classes 4.8, 5.8, 6.8 and 10.9, stainless steel bolts and aluminium bolts if the shear plane passes through the threaded portion

f_{ub} is the characteristic ultimate strength of the bolt material

A is the tensile stress area A_S of the bolt if the shear plane passes through the threaded portion of the bolt and the gross cross-section A if the shear plane passes through the unthreaded portion of the bolt.

Table 8.3 gives the tensile stress area and the design shear resistance for steel and stainless steel bolts if the shear plane passes through the threaded portion of the bolt and also the hole diameter according to EN 1090-3 for non-fitted bolts (normal clearances) and fitted bolts. Table 8.4 gives the gross cross-section area and the design shear resistance for steel and stainless steel bolts if the shear plane passes through the unthreaded portion of the bolt.

Bearing resistance

$$F_{b,Rd} = \frac{k_1 \alpha_b f_u d t}{\gamma_{M2}} \qquad (8.11)$$

where:

α_b is the smallest of α_d, f_{ub}/f_u or 1.0 \qquad (8.12)

Table 8.3. Design shear resistance $F_{v,Rd}$ (kN) for bolts if the shear plane passes through the threaded portion of the bolt ($\gamma_{M2} = 1.25$), and maximum hole diameters

Bolt	Tensile stress area, A_S	Steel			Stainless steel (EN 10088)		Maximum hole diameter	
		4.6	8.8	10.9	1.4301 $f_u = 50$	1.4436 $f_u = 80$	Standard (A, B, D, E)	Fitted bolts, (A, D, E)
M4	8.78	1.69	3.37	3.51	1.76	2.81	5	4.3
M5	14.2	2.73	5.45	5.68	2.84	4.54	6	5.3
M6	20.1	3.86	7.72	8.04	4.02	6.43	7	6.3
M8	36.6	7.03	14.1	14.6	7.32	11.7	9	8.3
M10	58.0	11.1	22.3	23.2	11.6	18.6	11	10.3
M12	84.3	16.2	32.4	33.7	16.9	27.0	13	12.3
M14	115	22.1	44.2	46.0	23.0	36.8	15	14.3
M16	157	30.1	60.3	62.8	31.4	50.2	17	16.3

Table 8.4. Design shear resistance $F_{v,Rd}$ (kN) for bolts if the shear plane passes through the unthreaded portion of the bolt ($\gamma_{M2} = 1.25$)

Bolt	Shank area, A	Steel			Stainless steel (EN 10088)	
		4.6	8.8	10.9	1.4301 $f_u = 50$	1.4436 $f_u = 80$
M4	12.6	2.41	4.83	5.03	2.51	4.02
M5	19.6	3.77	7.54	7.85	3.93	6.28
M6	28.3	5.43	10.9	11.3	5.65	9.05
M8	50.3	9.65	19.3	20.1	10.1	16.1
M10	78.5	15.1	30.2	31.4	15.7	25.1
M12	113	21.7	43.4	45.2	22.6	36.2
M14	154	29.6	59.1	61.6	30.8	49.3
M16	201	38.6	77.2	80.4	40.2	64.3

- in the direction of the load transfer:

$$\alpha_d = \frac{e_1}{3d_0} \qquad \text{for end bolts} \tag{8.13}$$

$$\alpha_d = \frac{p_1}{3d_0} - \frac{1}{4} \qquad \text{for inner bolts} \tag{8.14}$$

- perpendicular to the direction of the load transfer:

$$k_1 = \min\left(2.8\frac{e_2}{d_0} - 1.7; 2.5\right) \qquad \text{for edge bolts} \tag{8.15}$$

$$k_1 = \min\left(1.4\frac{p_2}{d_0} - 1.7; 2.5\right) \qquad \text{for inner bolts} \tag{8.16}$$

where:

f_u	is the characteristic ultimate strength of the material of the connected parts
f_{ub}	is the characteristic ultimate strength of the bolt material
d	is the bolt diameter
d_0	is the hole diameter.

The symbols e_1, e_2, p_1, p_2 are defined in Figure 8.1.

Tension resistance

The design tension resistance of the bolt–plate assembly $B_{t,Rd}$ should be taken as the smaller of the design tension resistance $F_{t,Rd}$ of the bolt given by *expression 8.17*, and the design punching shear resistance of the bolt head and the nut in the plate $B_{p,Rd}$ obtained from *expression 8.19*:

$$F_{t,Rd} = \frac{k_2 f_{ub} A_s}{\gamma_{M2}} \tag{8.17}$$

where:

k_2	$= 0.9$ for steel bolts
k_2	$= 0.50$ for aluminium bolts
k_2	$= 0.63$ for countersunk steel bolts
A_s	is the tensile stress area of the bolt.

Table 8.5 give the tensile stress area and the design tension resistance for steel and stainless steel bolts.

Table 8.5. Design tension resistance $F_{t,Rd}$ (kN) for bolts ($\gamma_{M2} = 1.25$)

Bolt	Tensile stress area, A_S	Steel			Stainless steel (EN 10088)	
		4.6	8.8	10.9	1.4301 $f_u = 50$	1.4436 $f_u = 80$
M4	8.78	2.53	5.06	6.32	3.16	5.06
M5	14.2	4.09	8.18	10.2	5.11	8.18
M6	20.1	5.79	11.6	14.5	7.24	11.6
M8	36.6	10.5	21.1	26.4	13.2	21.1
M10	58.0	16.7	33.4	41.8	20.9	33.4
M12	84.3	24.3	48.6	60.7	30.3	48.6
M14	115	33.1	66.2	82.8	41.4	66.2
M16	157	45.2	90.4	113	56.5	90.4

Punching shear resistance

$$B_{p,Rd} = 0.6\pi d_m t_p f_u / \gamma_{M2} \tag{8.19}$$

where:

d_m is the mean of the across points and across flats dimensions of the bolt head or the nut or, if washers are used, the outer diameter of the washer, whichever is smaller

t_p is the thickness of the plate under the bolt head or the nut

f_u is the characteristic ultimate strength of the member material.

Combined shear and tension

$$\frac{F_{v,Ed}}{F_{v,Rd}} + \frac{F_{t,Ed}}{1.4 F_{t,Rd}} \leq 1.0 \tag{8.20}$$

The design resistances for tension and for shear through the threaded portion given above are restricted to bolts with rolled threads. For cut threads, the relevant values should be reduced by multiplying them by a factor of 0.85. Furthermore, the values for the design shear resistance $F_{v,Rd}$ given in *expression 8.9* apply only where the bolts are used in holes with nominal clearances not exceeding those for standard holes as specified in EN 1090-3. See Table 8.3.

8.5.6   Design resistances of rivets

Clause C.4.2

The shear, bearing and tension resistances of solid aluminium rivets having domed heads in accordance with the requirements of *clause C.4.2* and *Table 3.4* are given in Table 8.5. As a general rule, the grip length of a rivet should not exceed 4.5*d* for hammer riveting and 6.5*d* for press riveting.

To avoid pull-out failure, a single rivet or one row of rivets should not be used in single lap joints between flats.

Bearing and punching shear resistance

For the bearing resistance and punching shear resistance, *expressions 8.11* to *8.16* and *8.19* for bolts apply to rivets as well.

Shear resistance and tension resistance

$$F_{v,Rd} = F_{t,Rd} = \frac{0.6 f_{ub} A_0}{\gamma_{M2}} \tag{8.10), (8.18}$$

where:

f_{ur} is the characteristic ultimate strength of the rivet material

A_0 is the cross-sectional area of the hole.

Figure 8.4. Minimum head dimensions of solid shaft rivets (no countersunk). (Reproduced from EN 1999-1-1 (*Figure C.1*), with permission from BSI)

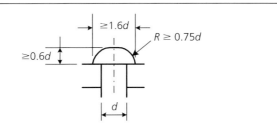

For solid rivets the head dimensions should be according to Figure 8.4 or greater on both sides.

8.5.7   Countersunk bolts and rivets
Special provisions are given in *clause 8.5.7* for countersunk steel bolts and rivets. No provisions are given for countersunk bolts and rivets made of aluminium.

Clause 8.5.7

8.5.8   Hollow rivets and rivets with break pull mandrel
For the design strength of hollow rivets and rivets with a break pull mandrel (blind rivets), see Section 8.5.15 in this guide (based on EN 1999-1-4).

8.5.9   High-strength bolts in slip-resistant connections
General
Provisions are given in *clause 8.5.9* for calculating the design resistance of slip-resistant connections using preloaded bolts. There are a number of specific provisions and limitations that apply to slip resistant connections in aluminium that can arise due to its comparatively low modulus and creep characteristics, or through lack of published research to enable design rules to be given:

Clause 8.5.9

- slip-resistant connections should only be used if the proof strength of the material of the connected parts is higher than 200 N/mm
- the bearing and shear capacity of the connection should always be sufficient at the ultimate limit state even if the connection is designed for slip resistance at the ultimate limit state (*clause 8.5.9.2*)
- holes should always be normal size – rules for oversize and slotted holes are not given (*clause 8.5.9.3(3)*)
- the slip factor decreases for connections between thinner elements (*clause 8.5.9.5*)
- it is necessary to retighten the bolts after a period (of at least 24 hours) (see EN 1090-3).

Clause 8.5.9.2

Clause 8.5.9.3(3)
Clause 8.5.9.5

Furthermore, it is pointed out that the effect of extreme temperature changes and/or long grip lengths that may cause a reduction or increase in the friction capacity due to the differential thermal expansion between aluminium and bolt steel cannot be ignored. However, no specific information is given in the code.

Resistance
The slip resistance can be utilised at the ultimate limit state or at the serviceability limit state only. However, at the ultimate limit state, the design shear force $F_{v,Ed}$ on a high-strength bolt should not exceed the lesser of

- the design shear resistance $F_{v,Rd}$
- the design bearing resistance $F_{b,Rd}$
- the tensile resistance of the member in the net section and in the gross cross-section.

Slip resistance/shear resistance
The design slip resistance of a preloaded high-strength bolt should be taken as

$$F_{s,Rd} = \frac{n\mu}{\gamma_{Ms}} F_{p,C} \qquad (8.21)$$

where:

$F_{p,C}$ is the preloading force (see below)
μ is the slip factor (see below)
n is the number of friction interfaces.

For bolts in standard nominal clearance holes, the partial safety factor for the slip resistance γ_{Ms} should be taken as $\gamma_{Ms,ult}$ for the ultimate limit state and $\gamma_{Ms,ser}$ for the serviceability limit state, where $\gamma_{Ms,ult}$ and $\gamma_{Ms,ser}$ are given in Section 8.1.1 of this guide.

If the slip factor μ is found by tests, the partial safety factor for the ultimate limit state may be reduced by 0.1.

Slotted or oversized holes are not covered by these clauses.

Preloading
For high-strength bolts of grades 8.8 or 10.9 with controlled tightening, the preloading force $F_{p,C}$ to be used in the design calculations should be taken as

$$F_{p,C} = 0.7 f_{ub} A_s \qquad (8.22)$$

Slip factor
The design value of the slip factor μ is dependent on the specified class of surface treatment. The value of μ for grit blasting to achieve a roughness value $R_a = 12.5$ (see EN ISO 1302 and EN ISO 4288) without surface protection treatments additional to grit blasting should be taken from Table 8.6.

Note that surface protection treatments applied before shot blasting may lead to lower slip factors.

The calculations for any other surface treatment or the use of higher slip factors should be based on specimens representative of the surfaces used in the structure using the procedure set out in EN 1090-3.

Combined tension and shear
If a slip-resistant connection is subjected to an applied tensile force $F_{t,Ed}$ in addition to the shear force $F_{v,Ed}$ tending to produce slip, the slip resistance per bolt should be taken as follows:

Category B, slip-resistant at the serviceability limit state:

$$F_{s,Rd,ser} = \frac{n\mu(F_{p,C} - 0.8F_{t,Ed,ser})}{\gamma_{Ms,ser}} \qquad (8.23)$$

Category C, slip resistant at the ultimate limit state:

$$F_{s,Rd} = \frac{n\mu(F_{p,C} - 0.8F_{t,Ed})}{\gamma_{Ms,ult}} \qquad (8.24)$$

Table 8.6. Slip factor of treated friction surfaces (data from EN 1999-1-1, *Table 8.6*)

Total joint thickness: mm	Slip factor, μ
$12 \le \sum t < 18$	0.27
$18 \le \sum t < 24$	0.33
$24 \le \sum t < 30$	0.37
$30 \le \sum t$	0.40

Figure 8.5. Prying forces (Q)

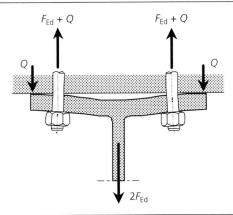

8.5.10 Prying forces

Where fasteners are required to carry an applied tensile force, they should be proportioned to also resist the additional force due to prying action, where this can occur (Figure 8.5).

The prying forces depend on the relative stiffness and geometrical proportions of the parts of the connection. A thick end plate results in small prying forces, whereas a thin end plate may result in large prying forces. If the effect of the prying force is taken advantage of in the design of the end plates, then the prying force should be allowed for in the design of the bolts. In *Annex B* (*'Equivalent T-stub in tension'* − see Section 9.2 in this guide), rules for the design of end plates, taking prying action into account, are given.

8.5.11 Long joints

In a lap joint, according to *clause 8.5.11*, the same bearing resistance in any particular direction should be assumed for each fastener up to a maximum length of max $L = 15d$, where d is the nominal diameter of the bolt or rivet (Figure 8.6). For $L > 15d$, the design shear resistance $F_{v,Rd}$ of all the fasteners should be reduced by multiplying it by a reduction factor β_{Lf}, given by

Clause 8.5.11

$$\beta_{Lf} = 1 - \frac{L_j - 15d}{200d} \qquad \text{but } 0.75 \leq \beta_{Lf} \leq 1.0 \qquad (8.25)$$

This provision does not apply where there is a uniform distribution of force transfer over the length of the joint (e.g. the transfer of shear force from the web of a section to the flange).

8.5.12 Single lap joints

Lap joints, in which the fasteners are in single shear, are found in truss gusset plates, in seams in plated structures and in secondary members. They are simple to fabricate and erect, but because of the inherent eccentricity of the load (Figure 8.7) they are subjected to local out-of plate bending, which also causes axial tension in the fastener. The greatest bending stresses

Figure 8.6. Lap joints. (Reproduced from EN 1999-1-1 (*Figure 8.10*), with permission from BSI)

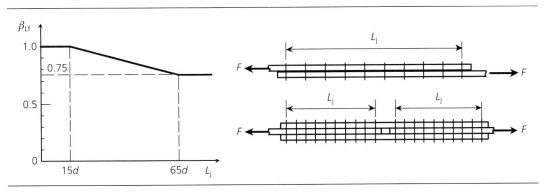

Figure 8.7. Single lap joint with one row of fasteners

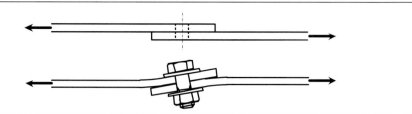

occur near the ends of the lap, and are most pronounced with single-fastener joints in short members.

Clause 8.5.12

Clause 8.5.12 states that in single lap joints of flats with one fastener or one row of fasteners (see Figure 8.7), bolts should be provided with washers under both the head and the nut, to avoid pull-out failure, unless the form of construction is such that rotation of the joint under load does not occur (e.g. in fully triangulated trusses). One single rivet or one row of rivets should not be used in such single lap joints; the same applies for countersunk bolts or rivets. Note, however, that, due to other reasons, bolts should generally be provided with washers under both the head and the nut for aluminium structures: see EN 1090-3, clause 8.2.6.

The bearing resistance $F_{b,Rd}$ determined in accordance with Section 8.5.5 in this guide should be limited to

$$F_{b,Rd} \leq 1.5 f_u dt / \gamma_{M2} \qquad (8.26)$$

8.5.13 Fasteners through packing
Where bolts or rivets transmit load in shear and bearing, and pass through packing of total thickness t_p greater than one-third of the nominal diameter d (Figure 8.8), the design shear resistance $F_{v.Rd}$ should be reduced by a factor β_p given by

$$\beta_p = \frac{9d}{8d + 3t_p} \quad \text{but } \beta_p \leq 1.0 \qquad (8.27)$$

For double shear connections with packing plates on both sides of the splice, t_p should be taken as the thickness of the thicker packing.

8.5.14 Pin connections
For connections made with pins (Figure 8.9), three cases may be recognised:

- if no rotation is required, the pin may be designed as if it were a single bolt
Clause 8.5.14
- if rotation is required, the procedures given in **clause 8.5.14** should be followed
- if the pin is to be designed as replaceable, a further limit on the contact bearing stress is
Clause 8.5.14
  applied (*expression 8.28* in **clause 8.5.14**).

Figure 8.12 gives geometric requirements for pin connections. Note that the provisions for edge and end distances differ from those given in *Table 8.2* for bolted and riveted connections.

Pins should not be loaded in single shear, so one of the members to be jointed should have a fork end or clevis (see Figure 8.9).

Figure 8.8. Fasteners through a packing plate

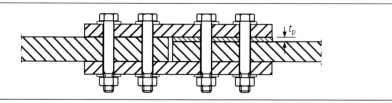

Figure 8.9. Pin connection

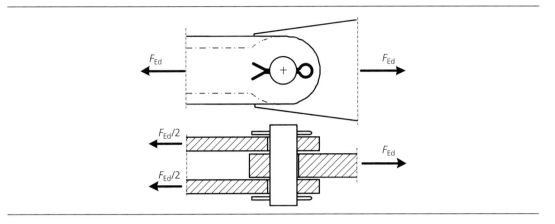

Table 8.7 lists the design requirements for pins in shear, bearing, bending, and combined shear and bending.

8.5.15   Self-tapping screw and blind rivets
Provisions for self-tapping screw and blind rivets are given in *Section 8* of EN 1999-1-4, '*Cold-formed structural sheeting*'.

General
The shear forces on individual mechanical fasteners in a joint may be assumed to be equal, provided that:

- the fasteners have sufficient ductility
- shear of the fastener is not the critical failure mode.

The partial factor for calculating the design resistances of mechanical fasteners should be taken as γ_{M3} according to Table 8.1.

Recommendations for the choice of fasteners for the risk of **corrosion** are given in *Annex B* of EN 1999-1-4, referred to in Section 10.10 of this guide.

Self-tapping screws may be penetrating, drilling or thread-forming (Figure 8.10). Drilling screws have some sort of drill tip. In carbon steel screws, the drill tip is part of the screw. Screws of austenitic steel have tips of special steel. Screws of martensitic stainless steel with a drill tip that is part of the screw may be encountered.

Riveting with a break pull mandrel (blind riveting) is a method of cold riveting carried out from one side of a joint. A riveting tool is used to withdraw the mandrel against the rivet so that the head of rivet is upset. There are various types of blind rivets: see the examples in Figure 8.11.

Distances and spacing
End distance, edge distance and spacing for fasteners according to Figure 8.1(a) should fulfil the following: $p_1 \le 30$ mm and $\le 4d$; $p_2 \le 20$ mm and $\le 2d$; $e_1 \le 20$ mm and $\le 2d$; and $e_2 \le 10$ mm and $\le 1.5d$.

Figure 8.10. Tips of (a) penetrating, (b) drilling and (c) thread-forming screws

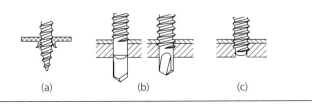

(a)   (b)   (c)

Figure 8.11. Different types of rivets with a break pull mandrel. (a) Open-end blind rivet. (b) Rivet with a captive mandrel head. (c) Closed-end blind rivet. (d) Rivet with a long break pull mandrel

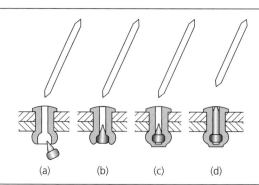

(a)   (b)   (c)   (d)

Combined shear and tension

For a fastener loaded in combined shear and tension, provided that $F_{p,Rd}$, $F_{v,Rd}$, $F_{b,Rd}$ and $F_{n,Rd}$ are determined according to Equations D8.1 to D8.7 below (**clauses 8.2** and **8.3** of EN 1999-1-4), the resistance of the fastener to combined shear and tension may be verified using

EN 1999-1-4
Clauses 8.2–8.3

$$\frac{F_{t,Ed}}{\min(F_{p,Rd}, F_{o,Rd})} + \frac{F_{v,Ed}}{\min(F_{b,Rd}, F_{n,Rd})} \leq 1 \qquad (8.1)$$

Bearing resistance

The bearing resistance if supporting members are of steel or aluminium is given by

$$F_{b,Rd} = \frac{2.5f_{u,min}}{\gamma_{M3}}\sqrt{t^3 d} \qquad \text{but } F_{b,Rd} \leq \frac{1.5f_{u,min}td}{\gamma_{M3}} \qquad \text{for } t_{sup}/t = 1 \qquad (D8.1)$$

$$F_{b,Rd} = \frac{1.5f_{u,min}td}{\gamma_{M3}} \qquad \text{for } t_{sup}/t \geq 2.5 \qquad (D8.2)$$

where t_{sup} is the thickness of supporting member.

For thickness $1.0 < t_{sup}/t < 2.5$, the bearing resistance $F_{b,Rd}$ may be obtained by linear interpolation.

The formulae apply to both screws and rivets.

Conditions:

- $f_{u,min} > 260\ \text{N/mm}^2$ should not be taken into account
- for $t > t_{sup}$ take $t = t_{sup}$
- drilling of the holes must be performed according to the recommendations of the manufacturer
- self-tapping and self-drilling screws should be of steel or stainless steel with a diameter $d \geq 5.5\ \text{mm}$
- rivets should have a diameter $2.6\ \text{mm} \leq d \leq 6.4\ \text{mm}$.

Net section resistance

$$F_{n,Rd} = \frac{A_{net}f_u}{\gamma_{M3}} \qquad (D8.3)$$

Shear resistance

EN 1999-1-4
Clause 8.2.2.3

Clause 8.2.2.3 of EN 1999-1-4 gives the shear resistance of aluminium blind rivets, which is different to solid rivets and should be taken as

$$F_{v,Rd} = \frac{38d^2}{\gamma_{M3}} \quad [\text{N}] \qquad \text{with } d \text{ in mm} \qquad (D8.4)$$

Table 8.7. Characteristic shear resistance $F_{v,Rk}$ N/screw for thread-forming screws

Outer diameter for threads: mm	Screw material	
	Hardened steel	Stainless steel
4.8	5 200	4 600
5.5	7 200	6 500
6.3	9 800	8 500
8.0	16 300	14 300

According to EN 1993-1-3 (cold-formed steel), the shear resistance of screws and rivets should be taken as

$$F_{v,Rd} = \frac{F_{v,Rk}}{\gamma_{M3}} \qquad \text{(D8.5)}$$

where $F_{v,Rk}$ is the characteristic shear strength of the fastener. For thread-forming screws the shear strength is obtained from Table 8.7 and for blind rivets from Table 8.8, where the strength of aluminium rivets is based on expression D8.4 and the other values are taken from the National Annexes for EN 1993-1-3 of the Nordic countries.

Tensile resistance
The tensile resistance of drilling screws may be taken as $F_{t,Rd} = 1.2F_{v,Rd}$.

Pull-through resistance
The pull-through resistance of joint in tension should be taken as

$$F_{p,Rd} = 6.1\alpha_L\alpha_E\alpha_M\sqrt{\frac{d_w}{22}}\frac{tf_u}{\gamma_{M3}} \quad \text{[N]} \qquad \text{with } t \text{ and } d_w \text{ in mm and } f_u \text{ in MPa} \qquad \text{(D8.6)}$$

where:

α_L is a factor that takes into account the tensile stress in the profile/plate:
if $f_u \geq 215$ MPa then $\alpha_L = 1.25 - L/6$ but $0.5 \leq \alpha_L \leq 1.0$ where L is span in m
if $f_u < 215$ MPa then $\alpha_L = 1$
at end supports without bending stresses and at connections at the upper flange (of profile sheeting) always $\alpha_L = 1$.
α_M is a factor that takes into account the type of washer:
$\alpha_M = 1.0$ for washers of steel or stainless steel
$\alpha_M = 0.8$ for washers of aluminium.
α_E is a factor that takes into account the position of the fastener in the fastened profile (Table 8.9).

The combination of correction factors is not necessary. The smallest value applies, which means that $\alpha_L\alpha_E\alpha_M = \min(\alpha_L; \alpha_E; \alpha_M)$ in expression D8.6.

Table 8.8. Characteristic shear resistance $F_{v,Rk}$ (N/rivet) for rivets with brake pull mandrel

Diameter: mm	Rivet material			
	Steel	Stainless steel	Monel	Aluminium
4.0	1600	2800	2400	600
4.8	2400	4200	3500	870
5.0	2600	4600	–	950
6.4	4400	–	6200	1500

Table 8.9. Correction factor α_E to take account of the location of the fasteners

Joint	For the flange in contact with the support					Without contact	
α_E	1.0	$b_u \leq 150$: 0.9 $b_u > 150$: 0.7	0.7 0.7	0.7	0.9	1.0	0.9

Reproduced from EN 1999-1-4 (*Table 8.3*), with permission from BSI.

Conditions:

- $t \leq 1.5$ mm
- $d_w \geq 14$ mm and thickness of the washer ≥ 1 mm
- the width of the adjacent flange of the sheet cross-section part ≤ 200 mm
- $d_w > 30$ mm and $f_u > 260$ MPa should not be taken into account
- at a depth of the profiled sheeting smaller than 25 mm, the pull-through-resistance should be reduced by 30%.

Pull-out resistance

$$F_{o,Rd} = 0.95 f_{u,sup} \sqrt{t_{sup}^3 d}\, \frac{1}{\gamma_{M3}}$$

(D8.7)

Conditions:

- self-tapping screws and self-drilling screws are of steel or stainless steel
- the diameter of the screws is 2.6 mm $\leq d \leq 6.5$ mm
- $t_{sup} > 6$ mm and $f_{u,sup} > 250$ MPa for aluminium should not be taken into account
- $t_{sup} > 5$ mm and $f_{u,sup} > 400$ MPa for steel should not be taken into account.

Blind rivets should not be used for joints in tension.

8.5.16 Screw ports and tracks for nut/bolt head

Screw ports and tracks for the nut/bolt head (Figure 8.12) are often used in extruded aluminium profiles. No resistance values are given in EN 1999-1-1. Some information on the resistance is given by Hellgren (1996) and Sapa Profiler (2009), valid for screws with a diameter between 3 mm and 7 mm.

Open screw port

An open screw port can be threaded for machine screws (Figure 8.12(a)), or can be used as is for self-tapping screws (Figure 8.12(b)). The thickness t of the material around the (standard) screw port should be $>0.38d_p$ where d_p is the diameter of the port, which in turn should be 0.9 times the diameter d of the screw.

The bearing resistance depends on the direction of the load. If the load is applied towards the closed side of the screw port, then the bearing resistance is the same as for a screw in solid

Figure 8.12. Open screw ports for (a) machine screws and (b) self-tapping screws, (c) closed screw port and (d) track for nut/bolt heads

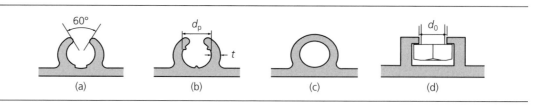

(a)   (b)   (c)   (d)

material, which means that the shear strength of the screw is decisive (see Table 8.3). If the force is applied towards the opening of the port, or perpendicular to the opening, the resistance depends on the material thickness.

Closed screw port
A closed screw port (Figure 8.12(c)) could be used where there are high strength requirements. The threaded grip length for metric fine threads should be larger than $3d_p$ for class 8.8 screws and larger than $4d_p$ for class 10.9 screws. The bearing and tensile resistance can be determined as for normal bolt holes, provided the thickness of the material is sufficient.

Tracks for nuts and bolt heads
Tracks for bolt heads and nuts (Figure 8.12(d)) can be used for the rapid joining of profiles or joining profiles to other components. The pull-out resistance depends on the shear area around the bolt head or nut.

The shear resistance of the load perpendicular to the track is about the same as for slotted holes.

The shear resistance of the load parallel to the track depends on the bolt torque and the variation of the load with time. Tracks for bolt heads and nuts should not be used in situations where the load can change direction.

Example 8.1: bolted connection
Calculate the design tensile resistance of the connection shown in Figure 8.13.

The dimensions are:
Connected part   width $b_{co} = 120$ mm, thickness $t_{co} = 12$ mm
Gusset plate   width at left bolt row $b_{gu} = 180$ mm, thickness $t_{gu} = 10$ mm
Plate material   EN AW-5754 H24, $f_u = 240$ MPa, $f_o = 160$ MPa
Distances and spacing $e_1 = e_2 = 30$ mm, $e_{2,gu} = 60$ mm, $p_1 = p_2 = 60$ mm
Bolt: 8.8, M16 $f_{ub} = 800$ MPa, shear plane passes through the unthreaded portion of the bolt
Hole diameter $d_0 = d + 1$ mm $= 17$ mm
Partial factors $\gamma_{M1} = 1.1$, $\gamma_{M2} = 1.25$

Shear strength of four M16 steel bolts

$$\alpha_v = 0.6, \ A = \pi d^2/4 = \pi \times 16^2/4 = 201 \text{ mm}$$

$$F_{v,Rd} = \alpha_v f_{ub} A / \gamma_{M2} = 0.6 \times 800 \times 201/1.25 = 77.1 \text{ kN}$$

(Table 8.4 gives same resistance)   (8.9)

Figure 8.13. Bolted connection

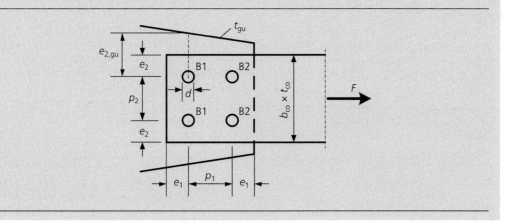

Table 8.10 Regular and minimum distances and spacings

Distance or spacing	Value	Regular	Minimum
End distance	$e_1 = 30$ mm	$2d_0 = 34$ mm	$1.2d_0 = 20.4$ mm
Edge distance	$e_2 = 30$ mm	$1.5d_0 = 25.5$ mm	$1.2d_0 = 20.4$ mm
Spacing	$p_1 = 60$ mm	$2.5d_0 = 42.5$ mm	$2.2d_0 = 37.4$ mm
Spacing	$p_2 = 60$ mm	$3d_0 = 51$ mm	$2.4d_0 = 40.8$ mm

For four bolts:

$$F_{v,Rd} = 309 \text{ kN}$$

Distances and spacing
Distances and spacing are larger than the regular values according to Table 8.2, except for the end distance, which is larger than minimum value (Table 8.10).

All distances and spacings are less than maximum values for aluminium exposed to the weather or other corrosive influences.

Bearing strength
The thicknesses of the connected part and the gusset plate are different. Furthermore, the edge distances are different for the two parts. Therefore it is not clear which bearing surface is critical.

The resistance is given by

$$F_{b,Rd} = k_1 \alpha_b f_u d t_{co}/\gamma_{M2} \tag{8.11}$$

Bolt B1 in the connected part:

End bolt:

$$\alpha_b = \min\left(\frac{e_1}{3d_0}, 1\right) = \min\left(\frac{30}{3 \times 17}, 1\right) = 0.588 \tag{8.13}$$

Edge bolt:

$$k_1 = \min\left(\frac{2.8e_2}{d_0} - 1.7, \ 2.5\right) = \min\left(\frac{2.8 \times 30}{17} - 1.7, \ 2.5\right) = 2.5 \tag{8.15}$$

Hence

$$F_{b,Rd,co} = 2.5 \times 0.588 \times 240 \times 16 \times 12/1.25 = 54.2 \text{ kN}$$

Bolt B1 in the gusset plate:

Inner bolt:

$$\alpha_b = \min\left(\frac{p_1}{3d_0} - \frac{1}{4}, \ 1\right) = \min\left(\frac{60}{3 \times 17} - \frac{1}{4}, \ 1\right) = 0.926 \tag{8.14}$$

Edge bolt:

$$k_1 = 2.5 \qquad \text{as above}$$

Hence

$$F_{b,Rd,gu} = 2.5 \times 0.926 \times 240 \times 16 \times 10/1.25 = 71.2 \text{ kN}$$

The resistance of bolt B1 is the smallest of the resistance in the connected part and the gusset plate.

Bolt B2 in the connected part:

Inner bolt:

$\alpha_b = 0.926$ as for B1 above

Edge bolt:

$k_1 = 2.5$ as $e_{2,gu} > e_2$

Hence

$F_{b,Rd,co} = 2.5 \times 0.926 \times 240 \times 16 \times 12/1.25 = 85.4$ kN

Bolt B2 in the gusset plate:

End bolt:

$\alpha_b = 0.588$ as for B1 above

Edge bolt:

$k_1 = 2.5$ as above

Hence

$F_{b,Rd,gu} = 2.5 \times 0.588 \times 240 \times 16 \times 10/1.25 = 45.2$ kN

The resistance of bolt B2 is the smallest of the resistance in the connected part and the gusset plate.

The sum of the four bolts is

$F_{b,Rd} = 2 \times 54.2 + 2 \times 45.2 = 199$ kN

Resistance in the net section
Connected part:

$A_{net} = b_{co}t_{co} - 2d_0t_{co} = 120 \times 12 - 2 \times 17 \times 12 = 1032$ mm^2

$N_{u,Rd} = 0.9A_{net}f_u/\gamma_{M2} = 0.9 \times 1032 \times 240/1.25 = 178$ kN   (6.19a)

Gusset plate:

$A_{net} = b_{gu}t_{gu} - 2d_0t_{gu} = 180 \times 10 - 2 \times 17 \times 10 = 1460$ mm^2

$N_{u,Rd} = 0.9 \times 1460 \times 240/1.25 = 252$ kN

Resistance of the gross cross-section
Connected part:

$N_{o,Rd} = A_g f_o/\gamma_{M1} = 120 \times 12 \times 160/1.1 = 210$ kN   (6.18)

Gusset plate:

$N_{o,Rd} = 180 \times 10 \times 160/1.1 = 232$ kN

Resulting resistance
The design resistance of the connection is 178 kN in the net section of the connected part.

8.6. Welded connections

8.6.1 General

Clause 8.6

Clause 8.6 deals with welded joints in aluminium. This clause differs from the remainder of Eurocode 9 in that it compares stresses rather than forces. The clause requires a check on the weld itself and also of the adjacent heat affected zone (HAZ). In practice, the majority of

Clause 6.5

checks in the HAZ are carried out in accordance with *clause 6.5* such that only local tension and shear at the fusion boundary need be considered in this clause.

The design resistance is given for welds made using MIG or TIG welding processes in accordance with the requirements of EN 1090-3. The resistance is applicable for predominately static loads.

There are several other welding processes commonly used in workshop practice such as laser welding, friction welding and friction stir welding. There are also some solid state welding processes, including explosion welding, ultrasonic welding, diffusion welding, and cold and hot pressure welding. These processes enable aluminium to be welded to a wide range of other metals. Eurocode 9 does not give design provisions for these welding processes.

8.6.2 Heat-affected zone, HAZ

The HAZ should be taken into account for the following alloys (Table 8.11):

- heat-treatable alloys in temper T4 and above (6xxx and 7xxx series)
- non-heat-treatable alloys in work-hardening condition (3xxx, 5xxx and 8xxx series).

Note that even small welds to connect an attachment to a main member may considerably reduce the resistance of the member due to the presence of a HAZ.

Normal and shear stresses in HAZ regions should satisfy *expression 8.42*:

$$\sqrt{\sigma_{haz,Ed}^2 + 3\tau_{haz,Ed}^2} \leq \frac{f_{u,haz}}{\gamma_{Mw}} \qquad (8.42)$$

where:

- for butt welds the design section is at the toe of the weld (the full cross-section for full-penetration welds, and the effective throat section t_e for partial penetration welds, see Figure 8.14)
- for fillet welds at the fusion boundary and at the toe of the weld.

Strength reduction in the HAZ can be compensated for by locally increasing the thickness. Differences in thickness can be levelled out and weld preparations can be incorporated using good extrusion designs. See Figure 8.15.

8.6.3 Design of welded connections

Clause 8.6.3

Clause 8.6.3 gives provision for the strength of the welds and the HAZ. The ductility of aluminium welds is generally less than the ductility of the parent material and the HAZ. It is therefore beneficial to slightly oversize welds to aid redistribution of any stress concentrations

Table 8.11. Reduction factors for the material in the HAZ for examples of alloys in extruded profiles and plates

Extruded profile			Sheet, strip and plate		
Alloy	Temper	Ultimate strength, $\rho_{u,haz}$	Alloy	Temper	Ultimate strength, $\rho_{u,haz}$
6082	T4	0.78	3005	H14	0.64
	T5	0.69		H16	0.56
	T6	0.64	5754	H14	0.63
7020	T6	0.80	6082	T6	0.60

Figure 8.14. Failure planes in the HAZ adjacent to a weld. (Reproduced from EN 1999-1-1 (*Figure 8.21*), with permission from BSI)

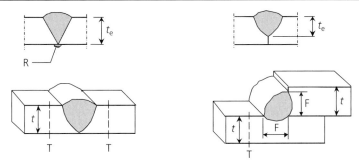

F, HAZ in the fusion boundary; T, HAZ in the toe of the weld, full cross-section; t_e, effective throat section; R, root bead

Figure 8.15. Welded connections. (a) Groove preparation, backing and support. (b) Differences in thickness. (c) Local increase in the thickness of the strength reduction zone. (d) Distance to a corner

(a)   (b)   (c)   (d)

through deformation of the adjacent material. This is particularly important if the distribution of load between welds is not based on the elastic distribution of stresses – see **clause 8.6.3.5**.

Clause 8.6.3.5

Characteristic strength of weld metal
Clause 8.6.3.1 gives the characteristic strength of weld metal for the most common combinations of parent metal and filler wire (Table 8.12). EN 1011-4 gives detailed recommendations for appropriate combinations, and it should be noted that it is sometimes better to use the filler metal with lower strength as it is less likely to give fabrication problems in joints that are highly restrained.

Clause 8.6.3.1

Design of butt welds
For full-penetration butt welds, the design resistance is taken as that of the weaker parts connected. The effective length should be taken as equal to the total weld length if run-on and run-off plates are used. Otherwise, the total length should be reduced by twice the thickness t. Partial penetration welds should not be used for primary load-bearing members (note that the wording of **clause 8.6.3.2.1** could imply that fillet welds should not be used, whereas it is intended only to deter partial-penetration welds).

Clause 8.6.3.2.1

Table 8.12. Characteristic strength values of weld metal f_w (MPa) (data from EN 1999-1-1, *Table 8.8*)

Filler metal	Alloy								
	3103	5052	5083	**5454** 5754 5049	**6060** 6063 3005 5005	**6005A** 6106	6061	6082 3004	7020
5356 5356A, 5056A 5556A, 5556B 5183, 5183A	–	170	240	220	160	180	190	210	260
4043A 4047A 3103	95	–	–	–	150	160	170	190	210

Figure 8.16. But weld subject to normal and shear stress

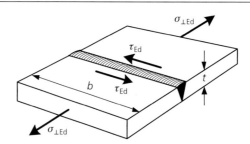

For the normal and shear stresses as shown in Figure 8.16 the following equation applies:

$$\sqrt{\sigma^2_{\perp ED} + 3\tau^2_{ED}} \le \frac{f_w}{\gamma_{Mw}}$$

(8.31)

where:

$\sigma_{\perp Ed}$ is the normal stress, perpendicular to the weld axis
τ_{Ed} is the shear stress, parallel to the weld axis
f_w is the characteristic strength of weld metal according to Table 8.13
γ_{Mw} is the partial safety factor for welded joints (see Section 8.1.1 in this guide).

Residual stresses and stresses not participating in the transfer of load need not be included when checking the resistance of a weld.

Design of fillet welds
A minimum length of eight times the throat thickness is required before the weld can be considered as load-carrying, where the throat thickness is the height of the largest triangle that can be inscribed within the weld. If the stress distribution along the length of the weld is not constant (see *Figure 8.16b* of EN 1999-1-1), and the length of the weld exceeds 100 times the throat thickness, the effective weld length $L_{w,eff}$ of longitudinal welds should be taken as

$$L_{w,eff} = \left(1.2 - 0.2\frac{L_w}{100a}\right)L_w \quad \text{for} \quad L_w \ge 100a$$

(8.32)

where:

L_w is the total length
a is the effective throat thickness.

For deep-penetration fillet welds, as defined by *Figure 8.17* of EN 1999-1-1, testing is necessary to demonstrate that the required degree of penetration can be achieved consistently.

Normal and shear stresses as shown in Figure 8.17 are assumed, in which:
$\sigma_\perp$ is the normal stress perpendicular to the throat section
$\sigma_\|$ is the normal stress parallel to the weld axis
$\tau_\perp$ is the shear stress acting on the throat section perpendicular to the weld axis
$\tau_\|$ is the shear stress acting on the throat section parallel to the weld axis.

The normal stress $\sigma_\|$, not participating in the load transfer, need not be included when checking the resistance of a fillet weld.

The design resistance of a fillet weld should fulfil

$$\sqrt{\sigma^2_{\perp Ed} + 3\left(\tau^2_{\perp Ed} + \tau^2_{\| Ed}\right)} \le \frac{f_w}{\gamma_{Mw}}$$

(8.33)

Figure 8.17. Stresses $\sigma_\perp$, $\tau_\perp$, $\sigma_\parallel$ and $\tau_\parallel$ acting on the throat section of a fillet weld. (Reproduced from EN 1999-1-1 (*Figure 8.18*), with permission from BSI)

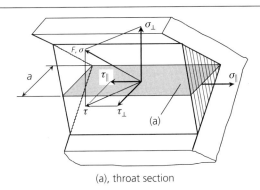

(a), throat section

Force at an angle to the weld axis

For a force F at an angle β to the weld axis, the stresses at the throat section of the components $F\sin\beta$ and $F\cos\beta$ will be (Figure 8.18)

$$\sigma_\perp = \frac{F\sin\beta}{aL\sqrt{2}} \qquad \tau_\perp = \frac{F\sin\beta}{aL\sqrt{2}} \qquad \tau_\parallel = \frac{F\cos\beta}{aL} \tag{D8.8}$$

Inserted into *Equation 8.33* we have

$$\sqrt{\left(\frac{F\sin\beta}{aL\sqrt{2}}\right)^2 + 3\left(\frac{F\sin\beta}{aL\sqrt{2}}\right)^2 + 3\left(\frac{F\cos\beta}{aL\sqrt{2}}\right)^2} \leq \frac{f_w}{\gamma_{Mw}} \tag{D8.9}$$

from which

$$\frac{F}{aL}\sqrt{2\sin^2\beta + 3\cos^2\beta} \leq \frac{f_w}{\gamma_{Mw}} \tag{D8.10}$$

The resistance for the force $F_{\beta Rd}$ at an angle $0° \leq \beta \leq 90°$ can now be derived as

$$F_{\beta Rd} = f(\beta)\frac{f_w aL}{\gamma_{Mw}} \tag{D8.11}$$

where

$$f(\beta) = \frac{1}{\sqrt{2\sin^2\beta + 3\cos^2\beta}} \tag{D8.12}$$

For longitudinal shear,

$$\beta = 0° \qquad F_{\parallel Rd} = \frac{1}{\sqrt{3}}\frac{f_w aL}{\gamma_{Mw}} \approx 0.577\frac{f_w aL}{\gamma_{Mw}} \tag{D8.13}$$

Figure 8.18. Forces and stresses on a fillet weld

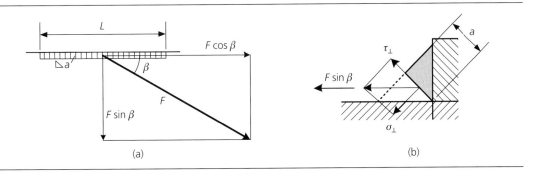

(a)   (b)

Figure 8.19. Double fillet welded joint loaded (a) parallel to and (b) perpendicular to the weld axis

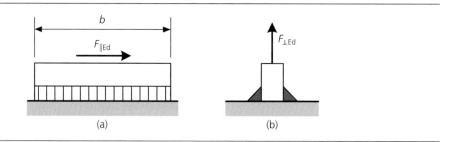

For a transverse load,

$$\beta = 90° \qquad F_{\perp Rd} = \frac{1}{\sqrt{2}} \frac{f_w a L}{\gamma_{Mw}} \approx 0.707 \frac{f_w a L}{\gamma_{Mw}} \tag{D8.14}$$

Clause 8.6.3.3

The required throat thickness a can be derived from expressions D8.13 and D8.14 (see *clause 8.6.3.3*). For a **double** fillet welded joint loaded parallel to the weld axis (Figure 8.19(a)), the throat thickness a should satisfy (note $0.5/0.577 \approx 0.866$ rounded to 0.85 in *expression 8.36* and D8.15)

$$a \geq 0.85 \frac{F_{\|Ed}}{b f_w / \gamma_{Mw}} \tag{D8.15}$$

where b is the width of the connected member.

For loading perpendicular to the weld axis (see Figure 8.19(b)), the throat thickness a should satisfy (note $0.5/0.707 \approx 0.707$ rounded to 0.7 in *expression 8.34* and D8.16)

$$a \geq 0.7 \frac{F_{\perp Ed}}{b f_w / \gamma_{Mw}} \tag{D8.16}$$

Design of connection with combined welds (weld groups)

Clause 8.6.3.5

For the design of connections with combined welds, two methods are given in *clause 8.6.3.5*:

- Method 1: the loads acting on the joint are distributed to the welds that are most suited to carry them.
- Method 2: the welds are designed for the stresses occurring in the adjacent parent metal of the different part of the joint.

8.7. Hybrid connections

Clause 8.7

Clause 8.7 allows for hybrid connections made up of different fastener types, such as a combination of welds with pre-loaded bolts that are designed for slip resistance at the ultimate limit state. As can be anticipated, combinations that do not have compatible deformation characteristics at the ultimate limit state are not allowed.

Clause 8.5.4

No provisions are given on the distribution of load between fasteners in this clause. The provisions of *clause 8.5.4* may, however, be used.

8.8. Adhesive-bonded connections

Recommendations for adhesive-bonded connections are given in *Annex M*. Note that testing is normally required to prove the design: see Section 9.12 in this guide.

8.9. Other joining methods

Rules for mechanical fasteners as given in EN 1999-1-4 are referred to in Section 8.5.15 of this guide. Other joining methods may be used, provided that appropriate tests are carried out.

Some other welding methods (including friction stir welding and laser welding) and mechanical fixing methods (including screw port connections and self-piercing rivets) are commonly used in aluminium workshops. Currently, there are no design rules for these in EN 1999-1-1, although it is hoped that subsequent revisions of the code will give suitable guidance.

Example 8.2: welded connection between a diagonal and a chord member

Calculate the tensile resistance of a welded connection of an angle diagonal to a chord member (Figure 8.20). The material in the diagonal and the chord is EN AW-6005A, with an ultimate strength of $f_u = 270$ MPa and $\rho_{u,haz} = 0.61$. The filler metal is 5356, with $f_w = 180$ MPa according to Table 8.13 (*Table 8.8* in EN 1999-1-1).

Figure 8.20. Connection of a diagonal to the chord of a built-up member

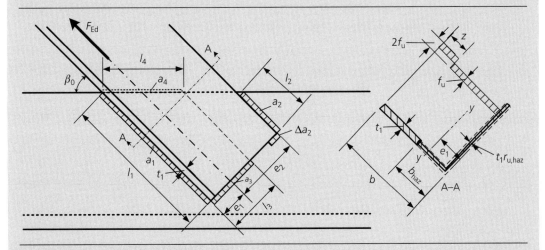

The angle between the diagonal and the chord is $\beta_0 = 42°$. The distance from the edge to the centre of gravity of the angle section 57 mm $\times$ 6 mm is $e_1 = 17$ mm, and the thickness $t_1 = 6$ mm.

The resistance of the four welds is derived with expression D8.11 in Table 8.13, where also the moment due to the eccentricity is calculated. This moment may be carried by an increase in welds 3 and 4:

$$\Delta l_3 = \frac{M_e}{(l_1 - 2e_1/\tan\beta_0)} \frac{\gamma_{Mw}}{f(0°)\, af_w} = \frac{62353}{90 - 2 \times 17/\tan 42°} \frac{1}{0.577 \times 4 \times 144} = 3.3\,\text{mm}$$

$$\Delta l_4 = \frac{f(42°)}{f(0°)} \Delta l_3 = \frac{0.626}{0.577} 3.3\,\text{mm} = 3.5\,\text{mm}$$

In practice, welds 3 and 4 are extended over the whole width of the angle section. Alternatively, weld 2 is completed with a weld Δa_2.

From Table 8.13, the sum of the resistance of the welds is

$$F_{Rd} = 54.6\,\text{kN}$$

Table 8.13. Weld data

Weld	l	a	β	e	f_w/γ_{Mw}	$f(\beta)$	F_{Rd}	M_e
1	90	3	0	17	144	0.577	22 447	381 605
2	32	3	0	−40	144	0.577	7 981	−319 252
3	34	3	90	0	144	0.707	10 386	0
4	51	3	42	0	144	0.626	13 791	0
						Sum:	54 605	62 353

Clause 6.1.6.3

The resistance of the HAZ is based on the net section area of the angle section, allowing for a HAZ all over the flange in the joint and b_{haz} in the other flange. The extent of the HAZ is $b_{haz} = 25$ mm according to Table 6.6 (*clause 6.1.6.3*) and the reduction factor $\rho_{u,haz} = 0.61$. The net section according to Section 6.2.3 in this guide is then

$$A_{net} = t(b - b_{haz}) + t(b + b_{haz} - t)\rho_{u,haz} = 6(57 - 25) + 6(57 + 25 - 6)0.61 = 470 \, mm^2$$

It is assumed that the tension force is acting in the plane of the joint. Then, a bending moment is acting on the angle section that is carried by the plastic distribution of stresses in the cross-section according to Figure 8.20. The compression part z is derived in such a way that the moment in the plane of the joint (middle of the angle leg) is zero (note $2tz$ on the right end side). Then,

$$\frac{t(b - t/2)^2}{2} - \frac{t(b_{haz} - t/2)^2}{2}\left(1 - \rho_{u,haz}\right) = 2tz(b - z/2)$$

from which

$$z = b - \sqrt{b^2 - \frac{(b - t/2)^2}{2} + \frac{(b_{haz} - t/2)^2\left(1 - \rho_{u,haz}\right)}{2}}$$

$$= 57 - \sqrt{57^2 - \frac{(57 - 3)^2}{2} + \frac{(25 - 3)^2(1 - 0.61)}{2}} = 13.6 \, mm$$

The resistance is

$$F_{Rd} = A_{net}\frac{f_u}{\gamma_{M2}} - tz\frac{2f_u}{\gamma_{M2}} = 470\frac{270}{1.25} - 6 \times 13.6\frac{2 \times 270}{1.25} = 66.3 \, kN$$

The design resistance is the resistance of the welds, which is

$$F_{Rd} = 54.6 \, kN$$

REFERENCES

Bulson PS (ed.) (1992) *Aluminium Structural Design Resent European Advances*. Elsevier, London.

Hellgren M (1996) Strength of bolt–channel and screw–groove joints in aluminium extrusions. Licentiate thesis, bulletin 24, Royal Institute of Technology, Department of Structural Engineering, Stockholm.

Mazzolani FM (ed.) (2002) *Aluminium Structural Design*. CISM Courses and Lectures No. 443. Springer, Wien.

Sapa Profiler (2009) *Sapa's Design Manual*. Sapa Profiler, Vetlanda.

Designers' Guide to Eurocode 9: Design of Aluminium Structures
ISBN 978-0-7277-5737-1

ICE Publishing: All rights reserved
http://dx.doi.org/10.1680/das.57371.161

Chapter 9
Annexes to EN 1999-1-1

In this chapter are gathered together the many annexes of EN 1999-1-1. The purpose is to provide an overview of the content in the annexes and to describe some important aspects of them.

Section numbers in this chapter do not relate to the code; however, equation numbers refer to EN 1999-1-1 unless specific to this guide.

Annex B is normative, the rest are informative.

9.1. Annex A – reliability differentiation
9.1.1 Introduction
Reliability management should follow the principles given in Eurocode 0, with execution in accordance with EN 1090-3. These require the designer to assess the consequences of failure and to choose relevant criteria for checking, execution, inspection and testing. Eurocode 0 introduces consequence classes and reliability classes, and these can be used to determine the parameters (execution class, service category and utilisation grade) required for the execution requirements of EN 1090-3.

Note that the consequence class (CC) and reliability class (RC) are related, such that CC1, CC2 and CC3 are associated with RC1, RC2 and RC3.

Annex A gives guidance on the choice of execution class, service category and utilisation grades to be used. In EN 1090-3 the default is that execution is carried out in accordance with execution class EXC2 unless otherwise stated.

Note: *Annex A* is informative. It is not recommended for use in the UK.

This section of the guide therefore summarises the guidance given in *Annex A* and also refers to the guidance given in PD 6702-1 (BSI, 2009a) for use in the UK.

9.1.2 Design supervision levels
Annex A simply refers to Eurocode 0 for design supervision levels. Eurocode 0 gives three levels of design supervision and checking as follows:

- **DSL3**, which is appropriate for structures of reliability class RC3. DSL3 requires extended supervision and the checking to be performed by a different organisation from that which prepared the design.
- **DSL2**, which is appropriate for structures of reliability class RC2. DSL2 requires normal supervision and checking by different persons from those originally responsible for the design.
- **DSL1**, which is appropriate for structures of reliability class RC1. DSL1 requires normal supervision and checking performed by the person who prepared the design.

In PD 6702-1 there is a recommendation that design supervision level DSL2 or DSL3 should be adopted when the default level of execution class EXC2 is selected.

9.1.3 Execution classes
EN 1999-1-1 adopts four levels of execution class (EXC1, EXC2, EXC3 and EXC4), where EXC4 has the most stringent requirements.

Table 9.1 Determination of execution class (data from EN 1999-1-1, *Table A.3*)

Consequence class		CC1		CC2		CC3	
Service category		SC1	SC2	SC1	SC2	SC1	SC2
Production category	PC1	EXC1	EXC1	EXC2	EXC3	EXC3[a]	EXC3[a]
	PC2	EXC1	EXC2	EXC2	EXC3	EXC3[a]	EXC4

[a] EXC4 should be applied to special structures or structures with extreme consequences of a structural failure also in the indicated categories as required by national provisions.

The execution class may apply to the whole structure or to parts of a structure. A structure may therefore have members or details that are of a different class to other members. This allows for the more stringent requirements for execution and inspection only to be applied for those details where it is necessary to do so.

Annex A gives guidance on choosing the execution class based on a matrix (Table 9.1) that considers the consequences of failure, the type of loading and the production category (whether the relevant components are welded).

PD 6702-1 does not use this matrix-based approach. The execution class provides a measure of the degree of assurance that the work has or will be carried out to the required quality. This is primarily differentiated by the extent of documented records as well as procedure and product testing. It does not define the quality requirement itself, which is dependent on the service category – see Section 9.1.4 of this guide. PD 6702-1 is designed to be used in conjunction with PD 6705-3 (BSI, 2009b).

PD 6702-1 recommends that EXC2 is used as the default class. It gives descriptive guidance of simple structures to indicate where EXC1 is appropriate, based on the type of structure, service category, material type and joining methods. It also gives descriptive guidance for circumstances where EXC3 is appropriate for complex structures where the consequences of failure are high. It should be noted that the execution requirements for EXC4 in EN 1090-3 are identical to those for EXC3 apart from the use of locknuts and the degree of the welding coordinator's knowledge.

9.1.4   Service category
Annex A recognises two service categories, SC1 applicable to structures subject to quasi-static actions and SC2 to structures that are subject to fatigue.

PD 6702-1 quantifies seven different service categories. The default category for structures subject to predominantly static loads and subject to high utilisation factors has the designation F20. Structures subject to predominantly static loads and but only to low utilisation factors has the designation F12. Structures subject to fatigue loads have designations ranging from F25 to F63, depending on the degree of fatigue utilisation determined in accordance with PD 6702-1. These different service categories can be used in conjunction with the inspection regimes and allowable imperfections given in PD 6705-3. The rationale behind the recommendations in PD 6702-1 and PD 6705-3 is to allow for the more stringent requirements for execution and inspection only to be applied where it is necessary to do so. Thus, a small amount of additional effort at the design stage can reduce execution costs.

9.1.5   Utilisation grades
Utilisation grades are used to determine the requirements for inspection and for acceptance criteria. For predominantly static loading, the utilisation grade is the ratio of the ultimate limit state design action divided by the design resistance as given by *Equation A1* in EN 1999-1-1:

$$U = \frac{E_k \gamma_F}{R_k / \gamma_M} \tag{A.1}$$

For fatigue loads, the utilisation grade is given by clauses in EN 1999-1-3 and PD 6702-1.

Figure 9.1. T-stub as basic component of other structural systems. (a) Unstiffened beam-to-column joint. (b) Stiffened beam-to-column joint. (Reproduced from EN 1999-1-1 (*Figure B.1*), with permission from BSI)

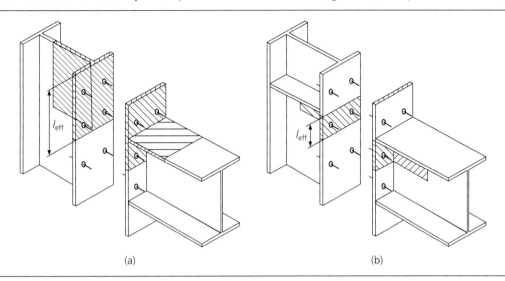

(a)   (b)

9.2.   Annex B – equivalent T stub in tension

The equivalent T stub is used in *Annex B* to model the resistance of the basic components of several structural systems. The possible modes of failure of the flange of an equivalent T stub may be assumed to be similar to those expected to occur in the basic component that it represents (Figure 9.1).

Generally, in bolted beam-to-column joints or beam splices it may be assumed that prying forces will develop unless the end plate is very thick or there is, for instance, a plate between the bolts, as in Figure 9.2(e), in which case the resistance is the lesser of flange failure and bolt failure. In cases where prying forces may develop, the tension resistance $F_{u,Rd}$ of a T stub flange should be taken as the smallest value for the four possible failure modes in Figure 9.2:

- Mode 1: flange failure by developing four hardening plastic hinges, two of which are at the web-to-flange connection (w) (with $\rho_{u,haz} \leq 1$) and two at the bolt location (b) (with $\rho_{u,haz} = 1$).
- Mode 2a: flange failure by developing two hardening plastic hinges with bolt forces at the elastic limit.
- Mode 2b: bolt failure with yielding of the flange at the elastic limit.
- Mode 3: bolt failure.

Formulae for the resistance of the different modes are given in *clause B.1(4)*, and are related in Example 9.1.

Clause B.1(4)

Figure 9.2. Failure modes of equivalent T stubs. (a) Mode 1. (b) Mode 2a. (c) Mode 2b. (d) Mode 3. (e) No prying action

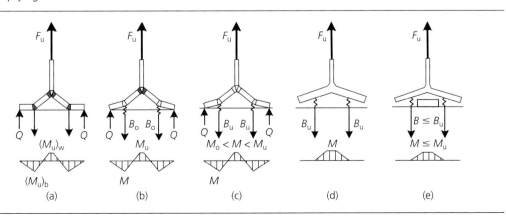

Methods for the determination of effective lengths l_{eff} for the individual bolt rows and bolt groups, for modelling basic components of a joint as equivalent T stub flanges, are given in *Table B.1* of *Annex B* for T stubs with unstiffened flanges and in *Table B.2* for T stubs with stiffened flanges. The effective length of an equivalent T stub is a notional length, and does not necessarily correspond to the physical length of the basic joint component that it represents.

Example 9.1: resistance of equivalent T-stub

Calculate the resistance of a T stub corresponding to a pair of bolts within c according to Figure 9.3.

Figure 9.3. Equivalent T stub

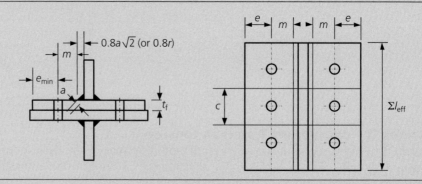

Material properties and measurements:

EN AW-6005A (*Table 3.2a*)	$f_o = 200$ MPa and $f_u = 250$ MPa
HAZ properties	$f_{o,haz} = 115$ MPa and $f_{u,haz} = 165$ MPa
Thickness of the flange plate	$t_f = 15$ mm
Lever arm	$m = 20$ mm
Edge distance	$e_{min} = 20$ mm
Bolt distance	$c = 30$ mm
Steel bolt 8.8	$d = 10$ mm

Clause B.1(4)

According to *clause B.1(4)*:

Ultimate strain:

$\varepsilon_u = 8\%$ from Table 3.2a

Elastic strain:

$$\varepsilon_o = \frac{f_o}{E} = \frac{200}{70\,000} = 0.00286$$

Strain relation:

$$\psi = \frac{\varepsilon_u - 1.5 \times \varepsilon_o}{1.5(\varepsilon_u - \varepsilon_o)} = \frac{0.08 - 1.5 \times 0.00286}{1.5(0.08 - 0.00286)} = 0.6543 \qquad (B.9)$$

Stress relation:

$$\frac{1}{k} = \frac{f_o}{f_u}\left(1 + \psi\frac{f_u - f_o}{f_o}\right) = \frac{200}{250}\left(1 + 0.6543\frac{250 - 200}{200}\right) = 0.931 \qquad (B.8)$$

Edge distance:

$$n = \min(e_{min}, 1.25m) = 20 \text{ mm}$$

8.8 steel bolt:

$$d = 10\,\text{mm} \quad d_0 = d + 1\,\text{mm} = 11\,\text{mm} \quad f_y = 640\,\text{MPa} \quad f_{ub} = 800\,\text{MPa}$$

Yield strength:

$$B_o = 0.9\frac{A_s f_y}{\gamma_{M2}} = 0.9\frac{58 \times 640}{1.25} = 26.7\,\text{kN} \tag{B.10}$$

Ultimate strength:

$$B_u = 0.9\frac{A_s f_{bu}}{\gamma_{M2}} = 0.9\frac{58 \times 800}{1.25} = 33.4\,\text{kN} \quad (= F_{t,Rd}) \tag{8.17}$$

Effective length 2 of the section at the edge of the weld:

$$l_{eff,2} = 30\,\text{mm}$$

Effective length 1 of the section through the bolt hole:

$$l_{eff,1} = l_{eff,2} - d_o = 30 - 11 = 19\,\text{mm}$$

Moment resistances in sections 1 and 2:

$$M_{u,1} = \frac{1}{4}t_f^2 \sum (l_{eff,1}f_u)\frac{1}{k}\frac{1}{\gamma_{M2}} = \frac{1}{4}15^2 \times 19 \times 250 \times 0.931\frac{1}{1.25} = 0.199\,\text{kN m} \tag{B.5}$$

$$M_{u,2} = \frac{1}{4}t_f^2 \sum (l_{eff,2}f_{u,haz})\frac{1}{k}\frac{1}{\gamma_{M2}} = \frac{1}{4}15^2 \times 30 \times 165 \times 0.931\frac{1}{1.25} = 0.207\,\text{kN m} \tag{B.6}$$

$$M_{o,2} = \frac{1}{4}t_f^2 \sum (l_{eff,2}f_{o,haz})\frac{1}{\gamma_{M1}} = \frac{1}{4}15^2 \times 30 \times 115\frac{1}{1.1} = 0.176\,\text{kN m} \tag{B.7}$$

If there are no welds in section 2, replace $f_{u,haz}$ with f_u, and $f_{o,haz}$ with f_o.

Mode 1 – flange failure by developing two hardening plastic hinges at the web-to-flange connection (w) $(=M_{u,2})$ and two at the bolt location (b) $(=M_{u,1})$:

$$F_{u,Rd} = \frac{2(M_{u,2})_w + 2(M_{u,1})_b}{m} - \frac{2 \times 207 + 2 \times 199}{20} - 40.6\,\text{kN} \tag{B.1}$$

Mode 2a – flange failure by developing two hardening plastic hinges with bolt forces at the elastic limit:

$$F_{u,Rd} = \frac{2M_{u,2} + n\sum B_o}{m + n} = \frac{2 \times 207 + 20 \times 2 \times 26.7}{20 + 20} = 37.1\,\text{kN} \tag{B.2}$$

Mode 2b – bolt failure with yielding of the flange at the elastic limit:

$$F_{u,Rd} = \frac{2M_{o,2} + n\sum B_u}{m + n} = \frac{2 \times 176 + 20 \times 2 \times 33.4}{20 + 20} = 42.2\,\text{kN} \tag{B.3}$$

Mode 3 – bolt failure:

$$F_{u,Rd} = \sum B_u = 2 \times 33.4 = 66.8\,\text{kN} \tag{B.4}$$

The design resistance is the smallest value of the four failure modes:

$$F_{u,Rd} = 38.5\,\text{kN} \quad \text{for mode 2a}$$

In Figure 9.4, curves for the resistance for the different failure modes are drawn for flange thicknesses varying from 5 mm to 30 mm. For small thicknesses, mode 1 gives the smallest resistance, and for increasing thickness, modes 2a, 2b and 3 govern. Q is the prying force.

Figure 9.4. Resistance of T stubs with thicknesses from 5 mm to 30 mm

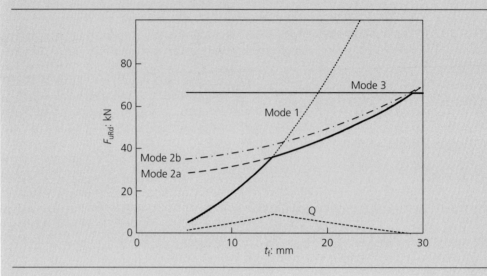

9.3. Annex C – material selection

The choice of a suitable aluminium material for any application is determined by a combination of factors; strength, durability, physical properties, weldability, formability and availability, both in the alloy and the particular form required.

Annex C gives descriptions of the 17 most commonly used wrought alloys and the six most commonly used cast alloys, indicating typical uses, durability and available forms. The descriptions also mention some of the comparative advantages and susceptibilities of the alloys, and advise when alternatives may be appropriate.

Tables C.1 and *C.2* (reproduced as Table 9.2 and 9.3 here) give a simple matrix that can be used in the initial evaluation of alloy properties.

Clause C.3.4

As noted in Chapters 1 and 3 of this guide, the design rules of EN 1999-1-1 have limited applicability for castings. *Annex C* (*clause C.3.4*) gives special rules that are recommended if castings are to be used.

Clause C.3.4.1

Clause C.3.4.1 covers design rules for castings. In essence, these require that the geometry should be such that applied actions do not give rise to buckling, and that design is carried out using a linear elastic analysis to compute the equivalent stress given by the following equation:

$$\sigma_{eq,Ed} = \sqrt{\sigma_{x,Ed}^2 + \sigma_{y,Ed}^2 + \sigma_{x,Ed}\sigma_{y,Ed} + 3\tau_{xy,Ed}^2} \qquad (C.1)$$

This equivalent stress should be compared with an allowable design strength taken as the lesser of $f_{oc}/\gamma_{Mo,c}$ and $f_{uc}/\gamma_{Mu,c}$. The partial factors can be defined in the National Annex, and it can be noted that the recommended values are higher than for wrought material.

Clause C.3.4.2

Clause C.3.4.2 lists specific quality and testing requirements for castings.

Clause C.4

Currently, there are no EN standards for aluminium bolts or for solid aluminium rivets. *Clause C.4* gives requirements that can be used to agree properties and dimensions with suppliers of aluminium bolts and solid rivets that can be used according to the design rules of EN 1999-1-1.

Table 9.2. Comparison of general characteristics and other properties for structural alloys (data from EN 1999-1-1, *Table C.1*)

Alloy: EN designation	Form and temper standardised for						Strength	Durability rating[a]	Weldability	Decorative anodising
	Sheet, strip and plate	Extruded products			Cold drawn products	Forgings				
		Bar/rod	Tube	Profile	Tube					
EN AW-3004	○	–	–	–	–		III/IV	A	I	I
EN AW-3005	○	–	–	–	–		III/IV	A	I	I
EN AW-3103	○	○	○	○	○		III/IV	A	I	II
EN AW-5005	○	○	○	○	○		III/IV	A	I	I
EN AW-5049	○	–	–	–	–		II/III	A	I	I/II
EN AW-5052	○	○	○ x)	○ x)	○		II/III	A	I	I/II
EN AW-5083	○	○	○ x)	○ x)	○	○	I/II	A	I	I/II
EN AW-5454	○	○	○ x)	○ x)	–		II/III	A	I	I/II
EN AW-5754	○	○	○ x)	○ x)	○	○	II/III	A	I	I/II
EN AW-6060	–	○	○	○	○		II/III	B	I	I
EN AW-6061	–	○	○	○	○		II/III	B	I	II/III
EN AW-6063	–	○	○	○	○		II/III	B	I	I/II
EN AW-6005A	–	○	○	○	–		II	B	I	II/III
EN AW-6106	–	–	–	○	–		II/III	B	I	I/II
EN AW-6082	○	○	○	○	○	○	I/II	B	I	II/III
EN AW-7020	○	○	○	○	○		I	C	I	II/III
EN AW-8011A	○	–	–	–	–	–	III/IV	B	II	III/IV

Key: ○ Standardised in a range of tempers; Availability of semi products from stock to be checked for each product and dimension; – Not standardised; x) Simple, solid sections only (seamless products over mandrel); I Excellent; II Good; III Fair; IV Poor.
Note: these indications are for guidance only, and each ranking is only applicable in the column concerned and may vary with temper.
[a] See *Table 3.1a*.

Table 9.3. Comparison of casting characteristics and other general properties (data taken from EN 1999-1-1, *Table C.2*)

Casting alloy: designation	Form of casting		Castability	Strength	Durability rating	Decorative anodising	Weldability
	Sand	Chill or permanent mould					
EN AC-42100		●	II	II/III	B	IV	II
EN AC-42200		●	II	II	B	IV	II
EN AC-43300	●	●	I	II	B	V	II
EN AC-43000		●	I/II	IV	B	V	II
EN AC-44200	●	●	I	IV	B	V	II
EN AC-51300	●	●	III	IV	A	I	II

Key: I Excellent; II Good; III Fair; IV Poor; V Not recommended; ● Indicates the casting method recommended for load bearing parts for each alloy.
Note 1: these indications are for guidance only, and each ranking is only applicable in the column concerned.
Note 2: the properties will vary with the condition of the casting.

9.4.   Annex D – corrosion and surface protection

Structures made of the aluminium alloys listed in *Section 3* of EN 1999-1-1 generally do not need any protective treatment to maintain structural integrity for typical lives of buildings and civil engineering structures in normal atmospheric conditions. Areas or environments that give conditions where protective treatment is likely to be required include:

Table 9.4. Recommendations for corrosion protection for various exposure conditions and durability ratings (data from EN 1999-1-1, *Table D.1*)

Alloy: durability rating	Material thickness: mm	Protection according to the exposure							
		Atmospheric						Immersed	
		Rural	Industrial/urban		Marine			Freshwater	Sea water
			Moderate	Severe	Non-industrial	Moderate	Severe		
A	All	0	0	(Pr)	0	0	(Pr)	0	(Pr)
B	<3	0	0	(Pr)	(Pr)	(Pr)	(Pr)	Pr	Pr
	≥3	0	0	0	0	0	(Pr)	(Pr)	Pr
C	All	0	0[b]	(Pr)[b]	0[b]	0[b]	(Pr)[b]	(Pr)[a]	NR

Key: 0 Normally no protection necessary; Pr Protection normally required except in special cases, see *clause D.3.2*; (Pr) The need for protection depends on if there are special conditions for the structure, see *clause D.3.2*. In case there is a need, it should be stated in the specification for the structure; NR Immersion in sea water is not recommended; [a] For 7020, protection is only required in a heat-affected zone (HAZ) if heat treatment is not applied after welding; [b] If heat treatment of 7020 after welding is not applied, the need to protect the HAZ should be checked with respect to conditions, see *clause D.3.2*.
Note: for the protection of sheet used in roofing and siding see prEN 508-2: 1996.

- structures to be used in severe industrial or polluted marine environments
- structures that will be subject to immersion in water
- parts of structures in contact with concrete or plaster
- parts of structures in contact with other metals
- parts of structures in contact with soil
- parts of structures in contact with certain species of timber.

Annex D gives a commentary on the corrosion and appearance of aluminium in various environments.

Table D.1 (reproduced here as Table 9.4), recommends when overall corrosion protection measures are needed in a range of different exposures for alloys of durability rating A, B or C (see *Tables 3.1, C.1* and *C.2*).

Table D.2 details additional protection that is recommended at bolted or riveted connections at metal-to-metal contacts (aluminium to aluminium, aluminium to steel and aluminium to stainless steel). It can be noted that in many applications no additional protection is needed despite contact with dissimilar metals.

Guidance is also given for the protection of aluminium surfaces that are in contact with concrete, masonry, plaster, timber, soils, and chemicals or insulating materials commonly used in the building industry.

9.5. Annex E – analytical models for stress–strain relationship

Annex E provides the models for the idealisation of the stress–strain relationship of aluminium alloys. Piecewise models, three-linear models and continuous models are presented. These models are conceived in order to account for the actual elastic-hardening behaviour. For materials as aluminium alloys, the Ramberg–Osgood model is often used to describe the stress–strain relationship in the form $\varepsilon = \varepsilon(\sigma)$:

$$\varepsilon = \frac{\sigma}{E} + 0.002\left(\frac{\sigma}{f_o}\right)^n \qquad (E.15)$$

Clause E.2.2.2(5)

According to *clause E.2.2.2(5)*, based on extensive tests, the following values may be assumed in the Ramberg–Osgood formula:

- In the **elastic range**:

$$n = n_e = \frac{\ln(0.000001/0.002)}{\ln(f_p/f_o)} \tag{E.16}$$

in which the proportional limit f_p only depends on the value of the 0.2% proof stress f_o,

$$f_p = f_o - 2\sqrt{10 f_o/(\text{N}/\text{mm}^2)} \qquad \text{if } f_o > 160 \,\text{N}/\text{mm}^2 \tag{E.17}$$

$$f_p = f_o/2 \qquad \text{if } f_o \leq 160 \,\text{N}/\text{mm}^2 \tag{E.18}$$

- In the **plastic range**:

$$n = n_p = \frac{\ln(0.002/\varepsilon_u)}{\ln(f_o/f_u)} \tag{E.19}$$

Furthermore, according to experimental data, the values of ε_u for the several alloys are, according to *clause E.3*, calculated using an analytical expression obtained by the interpolation of available results. This expression provides an upper-bound limit for the elongation at rupture:

Clause E.3

$$\varepsilon_u = 0.30 - 0.22 \frac{f_o/(\text{N}/\text{mm}^2)}{400} \qquad \text{if } f_o < 400 \,\text{N}/\text{mm}^2 \tag{E.20}$$

$$\varepsilon_u = 0.08 \qquad \text{if } f_o \geq 400 \,\text{N}/\text{mm}^2 \tag{E.21}$$

This formulation can be used to quantify the stress–strain model beyond the elastic limit for plastic analysis purposes, but it is not relevant for material ductility judgement as used in *Annexes F, G* and *H* (see Sections 9.5, 9.6 and 9.7 in this guide).

Example 9.2: value of coefficients in the Ramberg–Osgood formula

Derive the coefficients in the Ramberg–Osgood formula for the material in an extruded rod/bar (ER/B) made of EN AW-6082 T6 with thickness $t \leq 20$ mm.

According to *Table 3.2b*, $f_o = 250 \,\text{N}/\text{mm}^2$ and $f_u = 295 \,\text{N}/\text{mm}^2$.

In the **plastic range**, *expressions E.20* and *E.19* give

$$\varepsilon_u = 0.30 - 0.22 \frac{f_o/(\text{N}/\text{mm}^2)}{400} = 0.30 - 0.22\frac{250}{400} = 0.163 \tag{E.20}$$

$$n_p = \frac{\ln(0.002/\varepsilon_u)}{\ln(f_o/f_u)} = \frac{\ln(0.002/0.163)}{\ln(250/295)} = 26.6 \tag{E.19}$$

In the **elastic range**, *expressions E.17* and *E.16* give

$$f_p = f_o - 2\sqrt{10 f_o/(\text{N}/\text{mm}^2)} = 250 - 2\sqrt{10 \times 250} = 150 \,\text{N}/\text{mm}^2 \tag{E.17}$$

$$n_e = \frac{\ln(0.000001/0.002)}{\ln(f_p/f_o)} = \frac{\ln(0.000001/0.002)}{\ln(150/250)} = 14.9 \tag{E.16}$$

Stress–strain curves corresponding to Ramberg–Osgood's formula (*expression E.15*) with the exponents n_e and n_p are shown in Figure 9.5. The dotted curve is the best fit to a tension test at small strains, and the solid curve to large strains. Usually, the dotted curve should be used for stability calculations, and the solid curve for derivation of the plastic moment resistance.

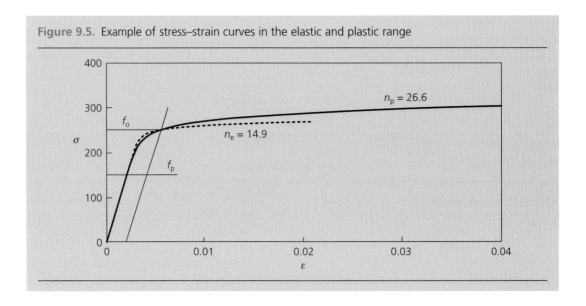

Figure 9.5. Example of stress–strain curves in the elastic and plastic range

9.6. Annex F – behaviour of cross-sections beyond the elastic limit

In *Annex F*, definitions of cross-section limit states and classifications of cross-sections are provided (see Section 6.1.4 in this guide). For cross-section class 1, a generalised shape factor α_1 (denoted α_{M1} for moments in *Annex F*) is given in *Table F.2* for the resistance beyond that corresponding to the geometrical shape factor $\alpha_0 = W_{pl}/W_{el}$:

$$\alpha_1 = \alpha_5 = 5 - \frac{(3.89 + 0.00190n)}{\alpha_0^{(0.270+0.0014n)}} \qquad \text{for 'brittle alloys' with } 4\% \leq \varepsilon_u < 8\% \qquad (D9.1)$$

$$\alpha_1 = \alpha_{10} = \alpha_0^{(0.21\log(1000n))} \times 10^{(0.0796-0.0809\log(n/10))} \qquad \text{for 'ductile alloys' with } \varepsilon_u \geq 8\% \qquad (D9.2)$$

where:

$n = n_p$ is the exponent of the Ramberg–Osgood law representing the material behaviour in the plastic range. Values in *Tables 3.2a* and *3.2b* may be used.

α_5, α_{10} are the generalised shape factors corresponding to ultimate curvature values $\chi_u = 5\chi_{el}$ and $\chi_u = 10\chi_{el}$, where χ_{el} is the elastic limit curvature.

α_0 is the geometrical shape factor.

For welded class 1 cross-sections, the following formula may be used:

$$\alpha_{1,haz} = \frac{\alpha_1}{\alpha_2} \frac{W_{pl,haz}}{W_{el}} \qquad (D9.3)$$

where α_1 (according to expression D9.1 or D9.2), and $\alpha_2 = W_{pl}/W_{el}$ are the shape factors for non-welded class 1 and 2 sections.

9.7. Annex G – rotation capacity

The provisions given in *Annex G* apply only to class 1 cross-sections, but may also be used for class 2 and 3 cross-sections, provided it is demonstrated that the rotation capacity is reached without local buckling of the section.

The stable part of the rotation capacity R is defined as the ratio of the plastic rotation at the collapse limit state $\theta_p = \theta_u - \theta_{el}$ to the limit elastic rotation θ_{el} (Figure 9.6):

$$R = \frac{\theta_p}{\theta_{el}} = \frac{\theta_u - \theta_{el}}{\theta_{el}} = \frac{\theta_u}{\theta_{el}} - 1 \qquad (G.5)$$

where θ_u is the maximum plastic rotation corresponding to the ultimate curvature χ_u.

Figure 9.6. Definition of rotation capacity. (Reproduced from EN 1999-1-1 (*Figure G.1*), with permission from BSI)

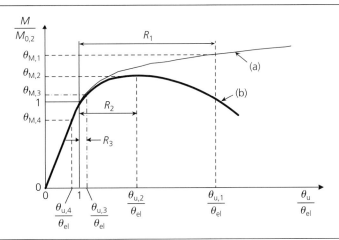

The following approximate formula may be used for the rotation capacity for class j cross-sections:

$$R_j = \alpha_j \left(1 + 2\frac{k\alpha_j^{m-1}}{m+1}\right) - 1 \qquad (G.6)$$

with m and k defined as

$$m = \frac{\ln\left[(10 - \alpha_{10})/(5 - \alpha_5)\right]}{\ln(\alpha_{10}/\alpha_5)} \qquad (G.3)$$

$$k = \frac{5 - \alpha_5}{\alpha_5^m}\left(=\frac{10 - \alpha_{10}}{\alpha_{10}^m}\right) \qquad (G.4)$$

where α_5 and α_{10} are the generalised shape factors according to expressions D9.1 and D9.2.

Example 9.3: shape factors and rotation capacity

Calculate the shape factors and rotation capacity for a rectangular cross-section of the material in Example 9.2 with the exponent of Ramberg–Osgood law $n_p = 29.6$.

$$\alpha_5 = 5 - \frac{(3.89 + 0.00190n)}{\alpha_0^{(0.270+0.0014n)}} = 5 - \frac{(3.89 + 0.00190 \times 29.6)}{\alpha_0^{(0.270+0.0014\times29.6)}} = 1.52 \qquad \text{for a 'brittle alloy'}$$

$$(D9.1)$$

$$\alpha_{10} = \alpha_0^{(0.21\log(1000n))} \times 10^{(0.0796-0.0809\log(n/10))} \qquad \text{for a 'ductile alloy'}$$

$$= \alpha_0^{(0.21\log(1000\times26.6))} \times 10^{(0.0796-0.0809\log(26.6/10))} = 1.62$$

$$(D9.2)$$

The shape factor is larger for the ductile alloy, as it should be:

$$m = \frac{\ln\left((10 - \alpha_{10})/(5 - \alpha_5)\right)}{\ln(\alpha_{10}/\alpha_5)} = \frac{\ln\left((10 - 1.62)/(5 - 1.52)\right)}{\ln(1.62/1.52)} = 14.3 \qquad (G.3)$$

$$k = \frac{5 - \alpha_5}{\alpha_5^m} = \frac{5 - 1.52}{1.52^{14.3}} = 0.00873 \left(k = \frac{10 - \alpha_{10}}{\alpha_{10}^m} = \frac{10 - 1.62}{1.62^{14.3}} = 0.00873\right) \qquad (G.4)$$

$$R_5 = \alpha_5 \left(1 + 2\frac{k\alpha_5^{m-1}}{m+1}\right) - 1 = 1.52\left(1 + 2\frac{0.00873 \times 1.52^{14.3-1}}{14.3+1}\right) - 1$$

$$= 0.976 \quad \text{if brittle alloy} \tag{G.6}$$

$$R_{10} = \alpha_{10} \left(1 + 2\frac{k\alpha_{10}^{m-1}}{m+1}\right) - 1 = 1.61\left(1 + 2\frac{0.00873 \times 1.62^{14.3-1}}{14.3+1}\right) - 1$$

$$= 1.72 \quad \text{if ductile alloy} \tag{G.6}$$

The rotation capacity of the ductile alloy is considerably larger than that of the brittle alloy.

9.8.   Annex H – plastic hinge method for continuous beams

In *Annex H*, provisions are given for the design of continuous beams with the plastic hinge method. These are:

- The cross-section belongs to class 1.
- The structural ductility is sufficient to enable the development of full plastic mechanisms.
- The plastic hinge method should not be used for members with transverse welds on the tension side of the member at the plastic hinge location.
- Adjacent to plastic hinge locations, any fastener holes in tension flange should satisfy

$$A_{f,\text{net}} 0.9 f_u / \gamma_{M2} \geq A_f f_o / \gamma_{M1} \tag{H.1}$$

  for a distance each way along the member from the plastic hinge location of not less than the greater of $2h_w$ and the distance to the adjacent point at which the moment in the member has fallen to 0.8 times the moment resistance at the point concerned.
- The rules are not applicable to beams where the cross-section varies along their length.

If applying the plastic hinge method to aluminium structures, both ductility and hardening behaviour of the alloy have to be taken into account. This leads to a correction factor η in the expression

$$M_{Rd} = \alpha_\xi \eta f_o W_{el} / \gamma_{M1} \tag{D9.4}$$

where $\alpha_\xi = \alpha_5$ or α_{10} according to expression D9.1 or D9.2, depending on the ductility of the alloy. The factor η is given by

$$\eta = \frac{1}{a - b/(n_p)^c} \quad \text{but } \eta \leq \frac{f_u / \gamma_{M2}}{f_o / \gamma_{M1}} \tag{D9.5}$$

where n_p is the Ramberg–Osgood exponent in the plastic range and the coefficients a, b and c are provided in Table 9.5.

If $\eta < 1$, the design load resistance should not be larger than evaluated through a linear elastic analysis.

Table 9.5. Values of the coefficients *a*, *b* and *c* in expression D9.5

Shape factor	Alloy	*a*	*b*	*c*
$\alpha_0 = 1.1\text{–}1.2$	Brittle	1.15	0.95	0.66
	Ductile	1.13	1.70	0.81
$\alpha_0 = 1.4\text{–}1.5$	Brittle	1.20	1.00	0.70
	Ductile	1.18	1.50	0.75

Example 9.4: bending moment resistance if the plastic hinge method is used

Calculate the bending moment resistance if the plastic hinge method is used for a small rectangular cross-section (height 20 mm, width 50 mm) of the ductile material in Examples 9.2 and 9.3.

The shape factor for a rectangular cross-section is 1.5, and the material is ductile. The coefficients in expression D9.5 are then found in Table 9.5 to be $a = 1.18$, $b = 1.50$ and $c = 0.75$, so

$$\eta = \frac{1}{a - b/(n_p)^c} = \frac{1}{1.18 - 1.50/26.6^{0.75}} = 0.951 \quad \text{but } \eta \leq \frac{f_u/\gamma_{M2}}{f_o/\gamma_{M1}} = \frac{295/1.25}{250/1.1} = 1.04$$

For the ductile alloy, $\alpha_{10} = 1.61$ according to Example 9.3, and the section module is

$$W_{el} = 50 \times 20^2/6 = 3333 \text{ mm}^3$$

The resistance is now, according to expression D9.4, for $f_o = 250$ MPa and Example 9.2,

$$M_{Rd} = \alpha_\xi \eta f_o W_{el}/\gamma_{M1} = 1.62 \times 0.951 \times 250 \times 3333/1.1 = 1.17 \times 10^6 \text{ N mm} = 1.17 \text{ kN m}$$

9.9. Annex I – lateral torsional buckling of beams and torsional or torsional flexural buckling of compression members

Annex I provides in *clause I.1* expressions for the elastic critical moment for beams and torsional or torsional flexural buckling of compression members for many loading conditions and cross-sections. Some of the expressions and values in the tables are used in Examples 6.12, 6.14, 6.17 and 9.5. In *clause I.2*, simplified formulae and coefficients for the slenderness for lateral torsional buckling are given for certain cross-sections. These are used in Example 6.12 for comparison with the 'exact' expressions in *clause I.1*.

Clause I.1

Clause I.2

Clause I.1

9.10. Annex J – properties of cross-sections

Annex J provides in *clause J.1* expressions for torsion constants, including factors for certain fillets and bulbs, in *clause J.2* the position of the shear centre and in *clause J.3* the warping constant. The beam in Example 6.14 has fillets that considerably increase the torsion stiffness.

Clause J.1
Clause J.2
Clause J.3

In *clause J.4*, a procedure is given where the cross-section is divided into rectangular parts defined by the coordinates of the ends. The procedure is illustrated in Example 9.5 (extruded profile) and Example 10.1 (cold-formed section). *Clause J.5* explains how to handle open sections with branches, and *clause J.6* the torsion constant for a cross-section with a closed (hollow) part.

Clause J.4

Clause J.5
Clause J.6

Example 9.5: lateral torsional buckling of an asymmetric beam with a stiffened flange

Calculate the elastic critical bending moment according to *clause I.1.2* for a 4 m beam loaded with a concentrated load on the top flange (Figure 9.7). There is no lateral or torsional restraint at the loading point. The compression flange has inclined lips (see the figure).

Clause I.1.2

To find the elastic critical buckling load, a number of cross-section constants need to be calculated. The procedures in *clauses J.4* and *J.5* are used, as the cross-section is not covered elsewhere. As the cross-section is asymmetric, the non-symmetry factor according to *expression J.27* is also needed. The calculations of the cross-section constants are made in a spreadsheet program, and the results are presented in Tables 9.6 to 9.8, where the expression numbering is given in the column head. The coordinates of the eight nodes are derived from Figure 9.7. All dimensions are in centimetres.

Clause J.4
Clause J.5

Figure 9.7. Cross-section and loading (dimensions in cm)

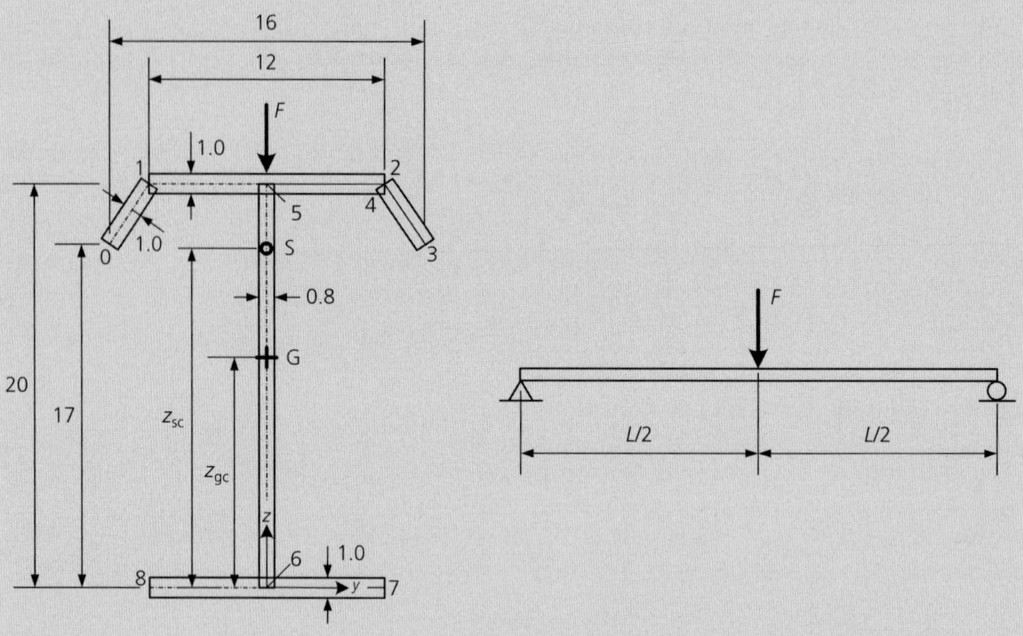

For example, for part 1 from node 0 to node 1:

$$A_1 = t_1\sqrt{(y_1 - y_0)^2 + (z_1 - z_0)^2} = 1.0 \times \sqrt{(-6 - (-8))^2 + (20 - 17)^2} = 3.61 \qquad (J.5)$$

$$S_{y,1} = (y_1 + y_0)A_1/2 = (20 + 17)3.61/2 = 66.6 \qquad (J.7)$$

$$I_{y0,1} = (z_1^2 + z_0^2 + z_1 z_0)A_1/3 = (20^2 + 17^2 + 20 \times 17)3.61/3 = 1237 \qquad (J.8)$$

From the sums in Table 9.6, the centre of gravity and the second moments of the area with respect to the centre of gravity are calculated:

$$z_{gc} = \frac{S_{y0}}{A} = \frac{533}{47.2} = 11.30 \qquad y_{gc} = \frac{S_{z0}}{A} = 0 \qquad (J.7)$$

Table 9.6. Calculation of the second moment of the area and the torsion constant

Node	Thickness	Coordinates		Area	Moments of area					Torsion
	t	y	z	A	S_y	I_y	S_z	I_z	I_{yz}	I_t
0		−8	17	(J.5)	(J.7)	(J.8)	(J.9)	(J.10)	(J.11)	(J.22)
1	1.0	−6	20	3.61	66.6	1237	−25	178	−465	1.20
2	1.0	6	20	12.0	240	4800	0	144	0	4.00
3	1.0	8	17	3.61	66.6	1237	25	178	465	1.20
4	0	6	20	0.0	0	0	0	0	0	0
5	0	0	20	0.0	0	0	0	0	0	0
6	0.8	0	0	16.0	160	2133	0	0	0	3.41
7	0	6	0	0.0	0	0	0	0	0	0
8	1.0	−6	0	12.0	0	0	0	144	0	4.00
			Sum	47.2	533	9407	0	644	0	13.8

Table 9.7. Calculation of sectorial constants

Node	Thickness	Coordinates		Area	Sectorial coordinates		Sectorial constants			
	t	y	z	A	ω_0	ω	I_ω	$I_{y\omega}$	$I_{z\omega}$	$I_{\omega\omega}$
0	0	−8	17	(J.5)	(J.15)	0	(J.16)	(J.17)	(J.18)	(J.19)
1	1	−6	20	3.61	−58	−58	−104.56	697	−1 987	4 043
2	1	6	20	12.0	−240	−298	−2 136	−2880	−42 720	437 808
3	1	8	17	3.61	−58	−356	−1 179	−8288	−21 760	386 549
4	0	6	20	0.0	58	−298	0	0	0	0
5	0	0	20	0.0	120	−178	0	0	0	0
6	0.8	0	0	16.0	0	−178	−2 848	0	−28 480	506 944
7	0	6	0	0.0	0	−178	0	0	0	0
8	1	−6	0	12.0	0	−178	−2 136	0	0	380 208
			Sum	47.2		−1722	−8 404	−10471	−94 946	1 715 552

$$I_y = I_{y0} - Az_{gc}^2 = 9407 - 47.2 \times 11.3^2 = 3380 \qquad (J.8)$$

$$I_z = I_{z0} - Ay_{gc}^2 = 644 - 47.2 \times 0^2 = 644 \qquad (J.10)$$

$$I_{yz} = I_{yz0} - \frac{S_{y0}S_{z0}}{A} = 0 - \frac{533 \times 0}{47.2} = 0 \qquad (J.11)$$

To calculate the sectorial constants, the sectorial coordinates are derived in Table 9.7.

For example, for part 3 from node 2 to node 3:

$$\omega_{0,3} = y_2z_3 - y_3z_2 = 6 \times 17 - 8 \times 20 = -58 \qquad (J.15)$$

$$\omega_3 = \omega_2 + \omega_{0,3} = -298 + (-58) = -356 \qquad (J.15)$$

The mean value ω_{mean} and sectorial constants are

$$\omega_{mean} = \frac{I_\omega}{A} = \frac{-8404}{47.2} = -178 \qquad (J.16)$$

Table 9.8. Calculation of sectorial constants

Node	ω	$\omega-\omega_{mean}$	$y-y_{gc}$	$z-z_{gc}$	y_c	z_c	$y_i - y_{i-1}$	$z_i - z_{i-1}$	[parenthesis]dA in	
0	0	178	−8.0	5.70	(J.29)	(J.29)			(J.28)	(J.27)
1	−58	120	−6.0	8.70	−7.00	7.20	2.0	3.0	−2 564	2 661
2	−298	−120	6.0	8.70	0	8.70	12.0	0	0	9 160
3	−356	−178	8.0	5.70	7.00	7.20	2.0	−3.0	2 564	2 661
4	−298	−120	6.0	8.70	7.00	7.20	−2.0	3.0	0	0
5	−178	0.0	0.0	8.70	3.00	8.70	−6.0	0	0	0
6	−178	0.0	0.0	−11.30	0	−1.30	0	−20.0	0	−2 112
7	−178	0.0	6.0	−11.30	3.00	−11.30	6.0	0	0	0
8	−178	0.0	−6.0	−11.30	0	−11.30	−12.0	0	0	−18 934
								Sum	0	−6 565

$$I_{y\omega} = I_{y\omega 0} - \frac{s_{z0}I_\omega}{A} = -10\,471 - \frac{0 \times (-8404)}{47.2} = -10\,471 \qquad (J.17)$$

$$I_{z\omega} = I_{z\omega 0} - \frac{s_{y0}I_\omega}{A} = -94\,946 - \frac{533 \times (-8404)}{47.2} = 0 \qquad (J.18)$$

$$I_{\omega\omega} = I_{\omega\omega 0} - \frac{I_\omega^2}{A} = 1\,715\,552 - \frac{-8404^2}{47.2} = 219\,715 \qquad (J.19)$$

The shear centre and the warping constant are as follows:

$$y_{sc} = \frac{I_{z\omega}I_z - I_{y\omega}I_{yz}}{I_yI_z - I_{yz}^2} = 0 \qquad (J.20)$$

$$z_{sc} = \frac{-I_{y\omega}I_y + I_{z\omega}I_{yz}}{I_yI_z - I_{yz}^2} = \frac{10\,471 \times 3380 + 0 \times 644}{3380 \times 644 - 0^2} = 16.26 \qquad (J.20)$$

$$I_w = I_{\omega\omega} + z_{sc}I_{y\omega} - y_{sc}I_{z\omega} = 219\,715 + 16.26 \times (-10\,471) - 0 \times 0 = 49\,500 \qquad (J.21)$$

The distance between the shear centre and the centre of gravity is

$$z_s = z_{sc} - z_{gc} = 16.27 - 11.30 = 4.97 \qquad (y_s = 0) \qquad (J.25)$$

The non-symmetry factor z_j is calculated in the spreadsheet program, where y_c and z_c are the coordinates for the centre of the cross-section part with respect to the centre of gravity and *parenthesis* is the expression in the sum of *formulae J.28 and J.27*.

From Table 9.8 and *expression J.27*,

$$z_j = z_s - \frac{0.5}{I_y}\sum [parenthesis]\,dA = 4.97 - \frac{0.5}{3380} \times (-6565) = 5.92 \qquad (J.27)$$

$$y_j = 0 \qquad (J.28)$$

Clause I.1.2

We now have all the cross-section constants needed to calculate the elastic critical bending moment according to *clause I.1.2* for the 4 m beam loaded with a concentrated load on the top flange. The coordinate of the load application point related to the shear centre is

$$z_g = z_a - z_s = 20 - 16.27 = 3.73 \qquad (D9.6)$$

For standard conditions of restraint at each end, $k_z = k_w = k_y = 1$, the non-dimensional parameters in *expression I.3* are for $E = 70 \times 10^5$ N/cm^2 and $G = E/2.6 = 27 \times 10^5$ N/cm^2:

$$\kappa_{wt} = \frac{\pi}{k_w L}\sqrt{\frac{EI_w}{GI_t}} = \frac{\pi}{1 \times 400}\sqrt{\frac{2.6 \times 49\,599}{13.8}} = 0.758$$

$$\zeta_g = \frac{\pi z_g}{k_z L}\sqrt{\frac{EI_z}{GI_t}} = \frac{\pi \times 3.73}{1 \times 400}\sqrt{\frac{2.6 \times 644}{13.8}} = 0.322$$

$$\zeta_j = \frac{\pi z_j}{k_z L}\sqrt{\frac{EI_z}{GI_t}} = \frac{\pi \times 5.92}{1 \times 400}\sqrt{\frac{2.6 \times 644}{13.8}} = 0.511$$

The factors C_1, C_2 and C_3 for a concentrated load is found in *Table I.2*. In footnote 1:

$$C_1 = C_{1,0} + (C_{1,1} - C_{1,0})\kappa_{wt} = 1.348 + (1.363 - 1.348)0.758 = 1.359$$

To find C_2 and C_3, the relation ψ_f according to *expression I.4b* is needed. For the actual cross-section $I_{ft} = 1 \times 12^3/12 = 144$ and $I_{fc} = I_z - I_{ft} = 644 - 144 = 500$, so

$$\psi_f = \frac{I_{fc} - I_{ft}}{I_z} = \frac{500 - 144}{644} = 0.553 \quad \text{which is within the limits } -0.9 \le \psi_f \le 0.9 \quad (I.4b)$$

so $C_2 = 0.553$ and $C_3 = 0.411$.

The relative non-dimensional critical moment according to *expression I.3*, and the elastic critical moment according to *expression I.2*, are, finally,

$$\mu_{cr} = \frac{C_1}{k_z}\left[\sqrt{1 + \kappa_{wt}^2 + (C_2\zeta_g - C_3\zeta_j)^2} - (C_2\zeta_g - C_3\zeta_j)\right]$$

$$= \frac{1.359}{1}$$

$$\times \left[\sqrt{1 + 0.758^2 + (0.553 \times 0.322 - 0.411 \times 0.511)^2} - (0.553 \times 0.322 - 0.411 \times 0.511)\right]$$

$$= 1.74 \tag{I.3}$$

$$M_{cr} = \mu_{cr}\frac{\pi\sqrt{EI_zGI_t}}{L}$$

$$= 1.74\frac{\pi\sqrt{70 \times 10^5 \times 644 \times 27 \times 10^5 \times 13.8}}{400} = 5.60 \times 10^6 \,\text{N cm} = 56.0\,\text{kN m} \tag{I.2}$$

9.11.   Annex K – shear lag effects in member design
Shear lag is briefly commented on in Section 6.2.2 of this guide.

9.12.   Annex L – classification of joints
General
According to *clause L.1* a **connection** is defined as the location where two members are interconnected, whereas a **joint** is the whole assembly of basic components that enabled members to be connected together. A joint may consist of one or more connections and parts of the joined members (e.g. a web panel in shear of a beam-to-column joint). With these definitions, what is in *Annex L* (and in the following) termed a connection could just as well be a joint.

Clause L.1

According to *Annex L*, connections may be classified in terms of their:

- (rotational) rigidity
- strength (moment resistance)
- (rotational) ductility.

These classifications are explained and exemplified in detail in the annex.

With respect to **rigidity**, connections may be classified as **rigid** or **semi-rigid**, depending on whether the initial stiffness corresponds to the connected member or not, regardless of strength and ductility.

With respect to **strength**, connections may be classified as **full strength** or **partial strength** connections, depending on whether the ultimate strength corresponds to the connected member or not, regardless of rigidity and ductility.

With respect to **ductility**, connections may be classified as **ductile**, **semi-ductile** or **brittle**, depending on whether the ductility of the connection is larger than or less than that of the connected members, regardless of rigidity and strength. Rotation limitations may be ignored

Figure 9.8. Example of moment–rotation curves

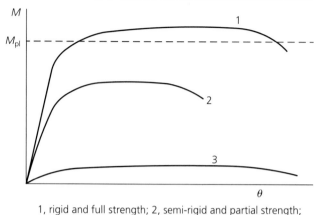

1, rigid and full strength; 2, semi-rigid and partial strength;
3, nominally pinned, all ductile

in the structural analysis if the connection is classified as ductile. For semi-ductile connections, rotation limitations must be considered in inelastic analysis, for brittle connections and also in elastic analysis.

In *Figure L.3*, several connection types are exemplified, depending on combinations of stiffness, strength and ductile properties, and in *Table L.1* they are shown with reference to the corresponding requirements for methods of global analysis.

Requirements for framing connections
With respect to the moment–curvature relationship, the connection types adopted in frame structures can be divided into:

- nominally pinned connections
- built-in connections.

Examples of the moment–rotation characteristics of joints are shown in Figure 9.8.

Connections defined as nominally pinned are incapable of transmitting significant moments and capable of accepting the resulting rotations under design loads.

Built-in connections allow for the transmission of bending moment between connected members, together with axial and shear forces. They can be classified according to rigidity and strength as follows:

- rigid connections
- semi rigid connections
- full strength connections
- partial strength connections.

Clause L.8.3

Descriptions of these connections are given in ***clause L.8.3***.

The rotation capacity of a connection may be demonstrated by experimental evidence: however, this is not required if using details that experience has proved have adequate properties in relation to the structural scheme.

Designers wishing to adopt the semi-continuous option should ensure that they are acquainted with the subject. This will require study of far more than just the provisions of Eurocode 9. Background texts include Mandara and Mazzolani (1995), but also texts concerning steel structures, such as Anderson (1996) and Faella *et al.* (2000). These texts explain the background to the concept of joint modelling necessary for the explicit inclusion of joint stiffness, partial strength properties and ductility when conducting a frame analysis.

9.13. Annex M – adhesive-bonded connections

Annex M covers adhesive-bonded connections, and notes in particular that bonding needs an expert technique and should be used with great care.

The design guidance in *Annex M* should only be used under the following conditions:

- the joint design is such that only shear forces have to be transmitted
- appropriate adhesives are applied
- the surface preparation procedures before bonding meet the specifications as required by the application.

The annex outlines many of the factors that have to be taken into account when using adhesive in connections. It notes that the configuration of the joint is crucial to avoid peel stresses and that knowledge of the adhesive strength in itself is not sufficient, although guidance on the characteristic shear strength of adhesives is given in *Table M.1*, together with the use of a high partial safety factor to be used with the strengths quoted in the table.

Prototype joint testing is recommended in ***clause M.2(5)***, and is essential, in the authors' view, to give a safe and economic design.

Clause M.2(5)

Annex M also refers to EN 1090-3 for execution. EN 1090-3, however, only requires that procedures are specified, without giving any detail. It can be observed that the correct performance of adhesive-bonded joints is only obtained if the recommendations of the adhesive manufacturer are followed rigorously.

In the UK, further requirements are recommended in PD 6705-3 that a work procedure should be prepared, based on the manufacturer's instructions, and should include the following:

- the full designation of the adhesive products
- the surface preparation of the parts
- the method of jigging and clamping the parts
- the method of preparing/mixing the adhesive products
- the tolerance limits on fit up
- the method of applying the adhesive to the parts
- restrictions on the environment (e.g. temperature, humidity)
- restrictions on time for the following operations:
  maximum shelf-life (storage before use)
  minimum mixing time
  maximum time between mixing and joint closure
  minimum clamping time
  minimum curing time prior to application of load
- inspection stages.

REFERENCES

Anderson D (ed.) (1996) *Semi-rigid Behaviour of Civil Engineering Structural Connections*. COST-C1, Brussels.

BSI (British Standards Institution) (2009a) PD6702-1:2009. Structural use of aluminium. Recommendations for the design of aluminium structures to BS EN 1999. BSI, London.

BSI (2009b) PD 6705-3. Structural use of steel and aluminium. Recommendations for the execution of aluminium structures to BS EN 1090-3. BSI, London.

Faella C, Piluso V and Rizzano G (2000) *Structural Steel Semi-rigid Connections*. CRC Press, Boca Raton, FL.

Mandara A and Mazzolani FM (1995) Behavioural aspects and ductility demand of aluminium alloy structures. *Proceedings of ICSAS '95*, Istanbul.

Designers' Guide to Eurocode 9: Design of Aluminium Structures
ISBN 978-0-7277-5737-1

ICE Publishing: All rights reserved
http://dx.doi.org/10.1680/das.57371.181

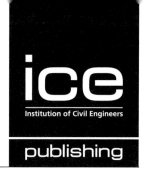

Chapter 10
Cold-formed structural sheeting

This chapter concerns the design of trapezoidal cold-formed structural sheeting, which is covered in EN 1999-1-4, 'Design of aluminium structures – Part 1-4: Cold-formed structural sheeting'. The purpose of this chapter is to provide an overview of the behavioural features of cold-formed structural components and to describe the important aspects of the code. The design of joints with mechanical fasteners as given in *Section 8* of EN 1999-1-4 is covered in Section 8.5.15 of this guide.

Section numbers in this chapter do not relate to the code; however, equation numbers refer to EN 1999-1-4 unless specific to this guide.

10.1. Introduction
Trapezoidal sheeting is used for the cladding of roofs and façades. They are usually designed for bending moment in the span and, if continuous, for the combination of the bending moment and the reaction force at inner supports. In addition, requirements concerning deflection at the serviceability limit state and the ability to walk on the sheeting during erection and maintenance may be decisive for the choice of thickness of the sheeting. EN 1999-1-4 does not cover the load arrangement for loads during execution and maintenance; however, the execution of aluminium structures made of cold-formed sheeting is covered in EN 1090-3.

The use of thin cold-formed material brings about a number of special design problems that are not generally encountered when using ordinary extruded or welded profiles. These include:

- rounded corners and the calculation of geometrical properties
- thickness and geometrical tolerances
- local buckling
- distortional buckling of flanges and webs with stiffeners
- shear lag
- flange curling
- web buckling due to transverse forces
- the durability of fasteners.

These effects, and the codified treatment, will be outlined in the remainder of this chapter.

10.2. Material properties, thickness, tolerances and durability
10.2.1 Material properties
Although all cold-formed operations involving plastic deformations result in changes to the basic material properties (essentially increasing the yield strength but with a corresponding reduction in the ductility), EN 1999-1-4 does not give any special credit for that. As characteristic values of 0.2% proof strength and ultimate strength, the values from the relevant product standard are adopted (see *Table 3.1* of EN 1999-1-4). It is assumed that the properties in compression are the same as those in tension.

10.2.2 Thickness and tolerances
According to *clause 3.2.2* of EN 1999-1-4, the nominal thickness of the sheeting exclusive of coatings should not be less than 0.5 mm. The nominal thickness should be used in the design if the negative deviation *dev* (%) is less than 5%. Otherwise, the design thickness *t* should be reduced to

Clause 3.2.2

$$t = t_{nom}(100 - dev)/95 \qquad (3.1)$$

For design by calculation, the maximum width-to-thickness ratios should fulfil $b/t \le 300$ for compression flanges and $s_w/t \le 0.5E/f_o$ for webs. Cross-sections with larger width-to-thickness ratios may be used, provided that their resistance and stiffness are verified by testing.

10.2.3 Durability
For basic requirements on the durability of aluminium structures, see *Section 4* of EN 1999-1-1. For cold-formed structural sheeting, special attention should be given to the risk of corrosion if the material in mechanical fasteners is such that electrochemical phenomena might produce conditions leading to corrosion. Recommendations for the choice of fasteners to avoid corrosion are given in *Annex B* of EN 1999-1-4 and Section 10.10 in this guide.

10.3. Rounded corners and the calculation of geometric properties
Cold-formed cross-sections contain rounded corners that make the calculation of geometric properties less straightforward than for the case of sharp corners. In such cross-sections, EN 1999-1-4 states in *clause 5.1(1)* that the notional flat width b_p, which is used as a basis for the calculation of the effective thickness, should be measured to the midpoint of adjacent corner cross-section parts, as shown in Figure 10.1.

Clause 5.1(1)

For small corner radii, the effect of the rounded corners is negligible and may be ignored. EN 1999-1-4 allows cross-section properties to be calculated based on an idealised cross-section that comprises flat parts concentrated along the mid-lines of the actual parts, provided $r \le 10t$ and $r \le 15b_p$, where r is the internal corner radius, t is the material thickness and b_p is the flat width of the cross-section part. Often, the radius fulfils the limits, so that sharp corners can be assumed. Example 10.1 shows the calculation of cross-section properties of trapezoidal sheeting based on the idealisations described. The influence of large rounded corners may approximately be taken into account by reducing the properties calculated on a cross-section with sharp corners according to *clause 5.1(4)* of EN 1999-1-4.

Clause 5.1(4)

10.4. Local buckling
As for extruded and welded profiles, an effective thickness approach is adopted: see Section 6.1.5 in this guide. In cases where the maximum compressive stress in a cross-section part is equal to the 0.2% proof strength ($\sigma_{com,Ed} = f_o/\gamma_{M1}$) then the reduction factor ρ should, according to *clause 5.5.2*, be obtained from

Clause 5.5.2

$$\rho = 1.0 \qquad \text{if } \bar{\lambda}_p \le \bar{\lambda}_{lim} \qquad (5.2a)$$

$$\rho = \alpha\left(1 - 0.22/\bar{\lambda}_p\right)/\bar{\lambda}_p \qquad \text{if } \bar{\lambda}_p > \bar{\lambda}_{lim} \qquad (5.2b)$$

in which the plate slenderness $\bar{\lambda}_p$ is given by

$$\bar{\lambda}_p = \sqrt{\frac{f_o}{\sigma_{cr}}} \equiv \frac{b_p}{t}\sqrt{\frac{12(1-\nu^2)f_o}{\pi^2 Ek_\sigma}} \cong 1.052\frac{b_p}{t}\sqrt{\frac{f_o}{Ek_\sigma}} \qquad (5.3)$$

Figure 10.1. Midpoint of corner to obtain notional widths of plane cross-section parts b_p allowing for corner radii. (Reproduced from EN 1999-1-4 (*Figure 5.1*), with permission from BSI)

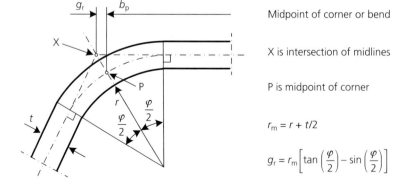

where k_σ is the relevant buckling factor from *Table 5.3* of EN 1999-1-4, $\bar{\lambda}_{\text{lim}} = 0.517$ and $\alpha = 0.9$.

If $\sigma_{\text{com,Ed}} < f_{\text{o}}/\gamma_{\text{M1}}$, then the reduction factor ρ may be determined using *expressions 5.2a* and *5.2b*, but replacing the plate slenderness $\bar{\lambda}_{\text{p}}$ by the reduced plate slenderness $\bar{\lambda}_{\text{p,red}}$ given by

$$\bar{\lambda}_{\text{p,red}} = \bar{\lambda}_{\text{p}}\sqrt{\frac{\sigma_{\text{com,Ed}}}{f_{\text{o}}/\gamma_{\text{M1}}}} \qquad (5.4)$$

For the calculation of the effective stiffness at serviceability limit states, see Section 7.2.4 in this guide, *clause 7.1(3)* of EN 1999-1-4 has the same provisions as for class 4 sections in EN 1999-1-1.

Clause 7.1(3)

In determining the effective thickness of a **flange** cross-section part subject to a stress gradient, the stress ratio ψ used in *Table 5.3* may be based on the properties of the gross cross-section.

In determining the effective thickness of a **web** cross-section part, the stress ratio ψ used in *Table 5.3* may be obtained using the effective area of the compression flange but the gross area of the web. Using the stress distribution based on the effective cross-section iteratively is optional.

10.5. Bending moment
10.5.1 General
The profile crest (top flange), profile trough (bottom flange) or webs of trapezoidal sheeting are often so slender that their resistance is reduced by local buckling (i.e. the cross-section belongs to class 4). However, classification of cross-sections as for extruded or welded cross-sections does not exist for cold-formed profiles, one of the reasons being that it is difficult to define slenderness limits for distortional buckling (see Section 10.5.3 in this guide). A cross-section may, however, be defined as belonging to class 4 if the effective cross-section is smaller than the gross cross-section. If local or distortional buckling does not reduce the cross-section (i.e. the effective section is the same as the gross section), in *clause 6.1.4* of EN 1999-1-4 there is an interpolation formula that allows for a certain degree of plastic resistance:

Clause 6.1.4

- If the effective section modulus W_{eff} is less than the gross elastic section modulus W_{el}:

$$M_{\text{c,Rd}} = W_{\text{eff}}f_{\text{o}}/\gamma_{\text{M1}} \qquad (6.4)$$

- If the effective section modulus W_{eff} is equal to the gross elastic section modulus W_{el}:

$$M_{\text{c,Rd}} = f_{\text{o}}\big(W_{\text{el}} + (W_{\text{pl}} - W_{\text{el}})4(1 - \lambda/\lambda_{\text{el}})\big)/\gamma_{\text{M1}} \qquad \text{but not more than } W_{\text{pl}}f_{\text{o}}/\gamma_{\text{M1}} \quad (6.5)$$

where λ is the slenderness of the cross-section part that corresponds to the largest value of $\lambda/\lambda_{\text{el}}$.

For the local buckling of internal cross-section parts (flanges and webs) $\lambda = \bar{\lambda}_{\text{p}}$ and $\lambda_{\text{el}} = \bar{\lambda}_{\text{lim}} = 0.517$, see Section 10.4 in this guide.

For stiffened cross-section parts $\lambda = \bar{\lambda}_{\text{s}}$ and $\lambda_{\text{el}} = 0.25$, see *Table 5.4* of EN 1999-1-4.

10.5.2 Sheeting without stiffeners
The calculation of the effective cross-section is carried out in a few stages (Figure 10.2), by obtaining:

1 the effective thickness of the flange in compression
2 the neutral axis GC1 of a cross-section with a reduced compression flange but an unreduced web
3 the effective thickness of the web part in compression
4 the new position GC2 of the neutral axis
5 the second moment of the area and the section modulus for the resulting effective cross-section
6 the bending moment resistance.

Figure 10.2. Calculation stages for trapezoidal sheeting

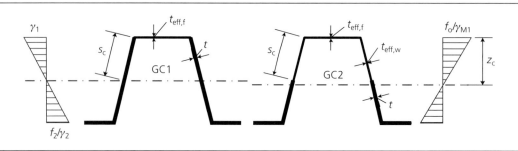

The effective thickness of the web may be based on a cross-section with a reduced compression flange and begin at the corresponding neutral axis GC1. It is, however, permitted to repeat stages 3 and 4 until SG1 and SG2 coincide, but this results in lower resistance.

10.5.3 Distortional buckling – calculation model for a flange with a stiffener

The strength of a flat cross-section part in compression can be increased by incorporating stiffeners in the shape of grooves or folds in the flanges and/or the webs. The behaviour of a stiffened flange is illustrated in Figure 10.3 (Höglund, 1980). The flat parts of the flange can buckle as an internal cross-section part with the buckling length approximately equal to the width of the flat part (Figure 10.3(a)).

The groove buckles in waves, with the half-wavelength about three to five times the width of the stiffened flange (Figure 10.3(b)). This is a form of distortional buckling that reduces the resistance of the groove itself and the adjacent flat parts.

The resistance is calculated in several stages:

1. the effective thicknesses of the flat parts are calculated on the assumption that the edges are pinned and that the groove is a rigid support
2. the groove and adjacent flat parts are treated as a member in compression, elastically braced by transverse plate strips as in Figure 10.3(c)
3. the cross-section area of the groove is reduced to an effective area that is a function of the distortional buckling load.

Figure 10.3. Model for buckling of a flange with one groove

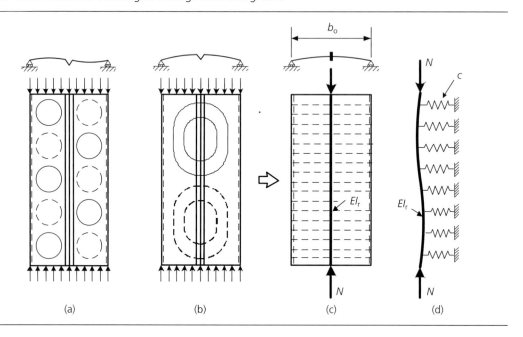

Figure 10.4. Plate strip in a flange with one concentric stiffener

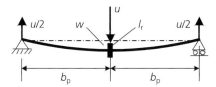

The buckling load is determined in the same way as for a member on an elastic foundation: see Figure 10.3(d). If the member is long, the buckling load is

$$N_{cr} = 2\sqrt{cEI} \tag{D10.1}$$

with the buckling half-length

$$l_{cr} = \pi\sqrt[4]{\frac{EI}{c}} \tag{D10.2}$$

The modulus of the elastic foundation (i.e. the spring constant per unit length) can be calculated as in the following simple example of a flange with one concentric stiffener.

The spring constant for a stiffener according to Figure 10.4 is determined from the deflection w of a plate strip across the flange due to a unit load u:

$$w = \frac{ub_p^3}{2 \times 3D} \tag{D10.3}$$

where D is the plate stiffness,

$$D = \frac{Et^3}{12(1 - v^2)} \quad (v = \text{Poisson's ratio} = 0.3) \tag{D10.4}$$

From this relationship, we obtain

$$c \equiv \frac{u}{w} = \frac{6D}{b_p^3} = \frac{6Et^3}{12(1-v^2)b_p^3} = \frac{Et^3}{2(1-v^2)}\frac{2^3}{b^3} \cong 4.4\frac{Et^3}{b^3} \tag{D10.5}$$

where $b = 2b_p$. The critical load for a stiffener in a long flange that buckles in several waves is

$$N_{cr} = 2\sqrt{cEI_s} = 2\sqrt{\frac{6DEI_s}{b_p^3}} \cong 4.2E\sqrt{\frac{I_s t^3}{8b_p^3}} \tag{D10.6}$$

where I_s is the second moment of the area of the stiffener. This formula can be used directly for a flange with a concentrated stiffener, such as one with an extruded aluminium profile (Figure 10.5(a)).

In a flange with grooves, the groove increases the deformation of the transverse strip. The cause of buckling is the compressive force in the groove and in adjacent flat parts. The load that acts on the plate strips in the transverse direction is distributed over a width that is equal to this effective part of the stiffener. For the sake of simplicity, two unit loads $u/2$ at the edges of the groove are used (Figure 10.5(b)), corresponding to the moment diagram (Figure 10.5(c)).

The deflection is increased with the first term in the following expression, where b_s is the developed width of the groove:

$$w_r = \frac{0.5ub_pb_s}{D}b_p + \frac{ub_p^3}{2 \times 3D} = \frac{ub_p^3}{6D}\left(1 + \frac{3b_s}{b_p}\right) \tag{D10.7}$$

Figure 10.5. Flange with one stiffener

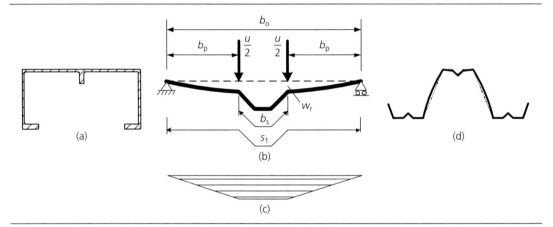

Substituting $c = u/w_r$ into $N_{cr} = 2\sqrt{cEI_s}$ gives

$$N_{cr} = 4.2E\sqrt{\frac{I_s t^3}{4b_p^2(2b_p + 3b_s)}} \tag{D10.8}$$

If N_{cr} is divided by the cross-sectional area of the groove, this gives the critical stress. *Expression 5.12* in EN 1999-1-4 is essentially the same but with a coefficient κ_w inserted, which takes into account the fact that the transverse plate strips are elastically restrained by the web, as indicated in Figure 10.5(d).

The derivation of the expressions for κ_w is not given here. It should, however, be explained that the restraint cannot be utilised if the length of the buckles in the flange is about the same as the length of the buckles in the web. This has been expressed by stating that full restraint can be assumed if the length of the flange buckles is more than 1.5 times the length of the web buckles. The length of the web buckles is, if it is assumed that the edges are pin jointed, about two-thirds of the depth of the web. The condition for full restraint is then that the length of the flange buckles should be double the depth s_w of the web. If the buckling length is smaller, the effect of restraint decreases according to a quadratic equation (*expression 5.14b*) in EN 1999-1-4.

10.5.4 One eccentric stiffener or two stiffeners
A similar derivation as above gives expression D10.9 for one eccentric stiffener (Figure 10.6(a)) and expression D10.10 for two symmetrically placed stiffeners (Figure 10.6(b)):

$$N_{cr} \cong 1.05E\frac{\sqrt{I_s t^3 b}}{b(b - b_k)} \tag{D10.9}$$

$$N_{cr} \cong 4.2E\sqrt{\frac{I_s t^3}{8b_k^2(3b_1 - 4b_k)}} \tag{D10.10}$$

Figure 10.6. Plate strip in a cross-section part with (a) an eccentric stiffener and (b) two symmetric stiffeners

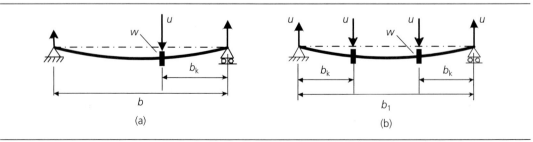

The eccentric stiffener applies, for instance, where there is a fold in a web. See *expression 5.23* in EN 1999-1-4, where κ_f takes account of the restraint by the flange. Since the compressed flange itself normally buckles over a length that is not much different from the length of the buckles in the web, κ_f is usually set $= 1.0$. Restraint by the flange in tension is approximately considered by using 0.9 times the depth of the web as the width b.

10.5.5   The effective area of a stiffener

The effective area of the stiffener including adjacent flat parts, which has been reduced due to local buckling, is reduced in a second step with respect to distortional buckling. The slenderness is determined in the usual way as

$$\overline{\lambda}_s = \sqrt{\frac{A_s f_o}{N_{cr}}} = \sqrt{\frac{f_o}{\sigma_{cr,s}}} \qquad (5.7)$$

The reduction factor χ_d for distortional buckling is according to *Table 5.4* in **clause 5.5.3.1** of EN 1999-1-4:

Clause 5.5.3.1

$$\chi_d = 1.155 - 0.62\overline{\lambda}_d \qquad \text{but } \leq 1.0 \quad \text{if } \overline{\lambda}_s \leq 1.04 \qquad (D10.11)$$

$$\chi_d = 0.53/\overline{\lambda}_s \qquad \qquad \text{if } \overline{\lambda}_s > 1.04 \qquad (D10.12)$$

The slenderness $\overline{\lambda}_s$ according to *expression 5.7* includes the area A_{ef} of the effective cross-section. This consists of the groove itself plus half the width of adjacent flat portions (Figure 10.7). The effective thickness of this portion depends on the yield stress or the largest compressive stress that occurs at the ultimate limit state.

Calculation of N_{cr} and σ_{cr} includes the second moment of the area I_s, which reflects the stiffness of the groove together with adjacent flat parts when buckling occur. This does not primarily depend on local buckling of the flat parts but, for instance, on the effect of local shear lag. The effective cross-section for the determination of I_s therefore consists of the groove itself plus $12t$ of adjacent flat portions (see Figure 10.7). Note that I_s is independent of the 0.2% proof strength f_o. Note also that the width $12t$ may be larger than the whole adjacent flat part. This does not, however, apply to the flat part between two grooves where the effective width cannot be larger than half that part (see Figure 10.7).

10.5.6   Sheeting with flange stiffeners and web stiffeners

Where there are stiffeners in both the flanges and the webs, the compressive force in the web fold also affects the critical load for the flange groove. Derivation of the critical load for this case is given in Höglund (1980) and is not referred to here. An approximate formula is given in **clause 5.5.4.4** of EN 1999-1-4.

Clause 5.5.4.4

Figure 10.7. Effective cross-section for the calculation of I_s and A_s for a compression flange with two stiffeners or one stiffener. (Reproduced from EN 1999-1-4 (*Figure 5.6*), with permission from BSI)

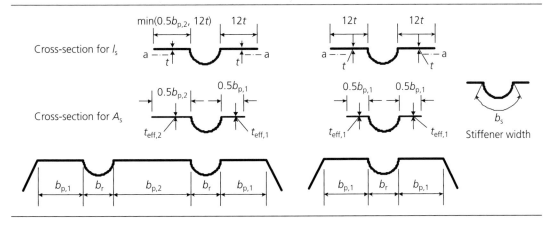

10.5.7 Plastic global analysis

If the slenderness of the compression parts is small enough (cross-section class 1), plastic global analysis may be used in calculating the moment redistribution in continuous sheeting. Note, however, that cross-section class 1 is not defined for trapezoidal sheeting; nevertheless, if there are no stiffeners, the classification for extruded profiles may be used.

There is also some residual resistance at an inner support beyond the maximum load in thin-walled sheeting, which means that moment redistribution is possible in many cases. This should be determined by tests: see *Annex A* of EN 1999-1-4.

Example 10.1: the bending moment resistance of trapezoidal sheeting with a stiffened flange

Calculate the effective cross-section and the bending moment resistance of the trapezoidal sheeting in Figure 10.8. The section properties are: top flange $b_c = 60$ mm, bottom flange $b_t = 90$ mm, section depth $h_w = 90$ mm, slant width of web $s_w = 96.57$ mm, pitch $b_{pi} = 220$ mm and plate thickness $t = 0.70$ mm. The internal corner radii fulfil the conditions $r \leq 10t$ and $r \leq 15b_p$, to adopt the idealised geometry. The 0.2% proof strength is 200 MPa, Young's modulus is 70 000 MPa and $\gamma_{M1} = 1.1$.

The groove dimensions are (see Figure 10.8): $b_r = 26$ mm, $h_r = 6$ mm, $c_r = 10$ mm and $s_r = 10$ mm, which gives $b_s = 30$ mm.

Calculation of the effective cross-section is carried out in several stages, by obtaining:

1 the effective thickness of the flat parts of the top flange in compression
2 the reduction in the stiffener area due to distortional buckling
3 the neutral axis of the cross-section with a reduced compression flange but an unreduced web
4 the effective thickness of the web part in compression
5 the new effective cross-section
6 the second moment of the area and the section modulus for the resulting effective cross-section
7 the bending moment resistance.

Flange curling and shear lag do not influence the resistance of this profile.

The cross-section constants are calculated using a spreadsheet program.

Figure 10.8. Idealised section of trapezoidal sheeting and the effective cross-section of a half pitch

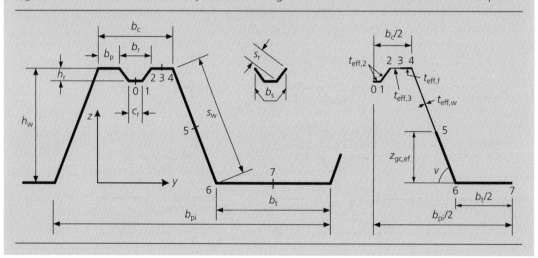

1. The effective thickness of the flat parts of the top flange in compression

Omitting calculation details, for the flat parts $b_p = (b_c - b_r)/2 = (60 - 26)/2 = 17$ mm and the buckling coefficient $k_\sigma = 4.0$, then the slenderness is $\bar{\lambda}_p = 0.683$, the reduction factor is $\rho = 0.893$ and the effective thickness is $t_{eff,f} = 0.6254$ mm.

2. The reduction of the stiffener area due to distortional buckling

The effective area of the stiffener is the groove plus half the adjacent flat parts on both sides of the groove:

$$A_s = b_s t + b_p t_{eff,f} = 30 \times 0.7 + 17 \times 0.6254 = 31.6 \, \text{mm} \tag{D10.13}$$

Second moment of the area of the stiffener = the groove $+ 12t$ on both sides of the groove:

$$I_s = c_r t h_r^2 + \frac{2s_r t h_r^2}{3} - \frac{(c_r t h_r + 2s_r t h_r/2)^2}{c_r t + 2s_r t + 2 \times 2t^2}$$

$$= 10 \times 0.7 \times 6^2 + \frac{2 \times 10 \times 0.7 \times 6^2}{3} - \frac{(10 \times 0.7 \times 6 + 10 \times 0.7 \times 6)^2}{10 \times 0.7 + 2 \times 10 \times 0.7 + 2 \times 12 \times 0.7^2}$$

$$= 205 \, \text{mm}^4 \tag{D10.14}$$

Developed width:

$$b_d = 2b_p + b_s = 2 \times 17 + 30 = 64 \, \text{mm} \tag{D10.15}$$

Buckling length and a factor to allow for restraint in the webs (*clause 5.5.4.2* in EN 1999-1-4):   *Clause 5.5.4.2*

$$l_b = 3.07 \sqrt[4]{I_s b_p^2 (2b_p + 3b_s)/t^3} = 3.07 \sqrt[4]{205 \times 17^2 (2 \times 17 + 3 \times 30)/0.7^3} = 209 \, \text{mm} \tag{5.15}$$

$$\kappa_{wo} = \sqrt{\frac{s_w + 2b_d}{s_w + 0.5b_d}} = \sqrt{\frac{96.6 + 2 \times 64}{96.6 + 0.5 \times 64}} = 1.322 \tag{5.16}$$

As

$$I_b/s_w = 288/96.6 > 2, \text{ then } \kappa_w = \kappa_{wo} = 1.322 \tag{5.14a}$$

Elastic critical buckling stress:

$$\sigma_{cr,s} = \frac{4.2\kappa_w E}{A_s} \sqrt{\frac{I_s t^3}{4b_p^2(2b_p + 3b_s)}} = \frac{4.2 \times 1.322 \times 70\,000}{31.6} \sqrt{\frac{205 \times 0.7^3}{4 \times 17^2(2 \times 17 + 3 \times 30)}}$$

$$= 272 \, \text{MPa} \tag{5.12}$$

Slenderness and the reduction factor from *Table 5.4* in EN 1999-1-4:

$$\bar{\lambda}_d = \sqrt{\frac{f_o}{\sigma_{cr,s}}} = \sqrt{\frac{200}{272}} = 0.858 \tag{5.14a}$$

$$\chi_d = 1.155 - 0.62\bar{\lambda}_d = 1.155 - 0.62 \times 0.858 = 0.623 \tag{D10.16}$$

Effective thickness of the groove:

$$t_{eff,2} = \chi_d t = 0.623 \times 0.7 = 0.436 \, \text{mm} \tag{D10.17}$$

Effective thickness of the flat parts adjacent to the groove:

$$t_{eff,3} = \chi_d \rho t = 0.623 \times 0.893 \times 0.7 = 0.3897 \, \text{mm} \tag{D10.18}$$

3. The neutral axis for a cross-section with a reduced compression flange but an unreduced web

The section properties for the gross cross-section and the section with a reduced compression flange are calculated using the formulae given in *Annex J* of EN 1999-1-1 (Table 10.1). The half-pitch is divided into seven parts between eight nodes according to Figure 10.8. Node 3 is added to define the middle of the flat part of the flange, and node 5 to define the compressed part of the web.

Centre of gravity and the second moment of the area of the gross cross-section (needed if the cross-section belongs to class < 4):

$$z_{gc,g} = \sum S_{y0,i} / \sum A_i = 5016/121.5 = 41.3 \text{ mm} \tag{J.7}$$

$$I_{y,g} = \sum I_{y0,i} - z_{gc,g}^2 \sum A_i = 356\,600 - 41.3^2 \times 121.5 = 149\,500 \text{ mm}^4 \tag{J.8}$$

4. The effective thickness of the web part in compression

Centre of gravity of the section with a reduced flange (from Table 10.1):

$$z_{gc,1} = \sum S_{y0,i} / \sum A_i = 4381/114.3 = 38.34 \text{ mm} \tag{J.7}$$

Stress distribution factor, buckling coefficient, slenderness and reduction factor:

$$\psi = \frac{-z_{gc,1}}{h_w - z_{gc,1}} = \frac{-38.34}{90 - 38.34} = -0.742$$

$$k_\sigma = 7.81 - 6.26\psi + 9.78\psi^2 = 7.81 + 6.26 \times 0.7422 + 9.78 \times 0.7422^2 = 17.84$$

for $0 > \psi \geq -1$ in Table 10.1

$$\bar{\lambda}_p = 1.052 \frac{s_w}{t} \sqrt{\frac{f_o}{Ek_\sigma}} = 1.052 \frac{96.6}{0.7} \sqrt{\frac{200}{70\,000 \times 17.84}} = 1.836 \tag{5.3}$$

$$\rho = \alpha \left(\frac{1}{\bar{\lambda}_p} - \frac{0.22}{\bar{\lambda}_p^2} \right) = 0.9 \left(\frac{1}{1.836} - \frac{0.22}{1.836^2} \right) = 0.431 \tag{5.2b}$$

Effective thickness of the compression part of the web:

$$t_{eff,w} = \rho t = 0.431 \times 0.7 = 0.3017 \text{ mm} \tag{D10.19}$$

Table 10.1. Spreadsheet for the gross cross-section and the section with a reduced compression flange

Node	t	Coordinates		Gross cross-section			Reduced compression flange			
		y	z	A	S_{y0}	I_{y0}	t_{eff}	A	S_{y0}	I_{y0}
0		30	84	(J.5)	(J.7)	(J.8)		(J.5)	(J.7)	(J.8)
1	0.7	35	84	3.5	294	24 696	0.4362	2.2	183	15 389
2	0.7	43	90	7.0	609	53 004	0.4362	4.4	379	33 029
3	0.7	51.5	90	6.0	536	48 195	0.3897	3.3	298	26 831
4	0.7	60	90	6.0	536	48 195	0.6254	5.3	478	43 059
5	0.7	77.5	45	33.8	2281	159 696	0.7	33.8	2281	159 696
6	0.7	95	0	33.8	760	22 814	0.7	33.8	760	22 814
7	0.7	140	0	31.5	0	0	0.7	31.5	0	0
			Sum	121.5	5016	356 600		114.3	4381	300 818

5. The effective cross-section

Node 5 is moved to the centre of gravity:

$$z_5 = z_{gc,1} = 38.34 \, \text{mm} \tag{D10.20}$$

$$y_5 = y_6 - \frac{z_5}{h_w}(y_6 - y_4) = 95 - \frac{38.34}{90}(95 - 60) = 80.09 \, \text{mm} \tag{D10.21}$$

6. The second moment of the area and the section modulus for the resulting effective cross-section

Using Table 10.2, the centre of gravity, second moment of area of the effective cross-section and section modulus are

$$z_{gc,ef} = \sum S_{y0,i} / \sum A_i = 2966/92.2 = 32.16 \, \text{mm} \tag{J.7}$$

$$I_{y,ef} = \sum I_{y0,i} - z_{gc,g}^2 \sum A_i = 205\,094 - 32.15^2 \times 92.2 = 109\,700 \, \text{mm}^4 \tag{J.8}$$

$$W_{ef} = I_{y,ef}/(h_w - z_{gc,ef}) = 109\,700/(90 - 32.16) = 1897 \, \text{mm}^3$$

7. The bending moment resistance

As the effective cross-section is less than the gross cross-section, the bending moment resistance per unit width is (based on *expression 6.4* in **clause 6.1.4.1**)

Clause 6.1.4.1

$$M_{Rd} = \frac{W_{ef}f_o}{b_{pi}\gamma_{M1}} = \frac{1897 \times 2 \times 200}{220 \times 1.1} = 3135 \, \text{N mm/mm} = 3.135 \, \text{kN m/m} \tag{D10.22}$$

8. The plastic section modulus

Although the plastic section modulus is not needed (as the effective cross-section is less than the gross cross-section in this example), a procedure to calculate the plastic section modulus of the cold-formed section will be presented. The cross-section (half-pitch) is divided into two parts with the same cross-section area. In this example, the plastic neutral axis crosses the web for

$$s_{pl} = \frac{A_g/2 - b_t t/2}{t} = \frac{121.5/2 - 90 \times 0.7/2}{0.7} = 41.79 \, \text{mm}$$

where the cross-section area A_g was found in Table 10.1. The corresponding coordinates are

$$z_{pl} = z_5 = s_{pl}\cos(v) = 41.79 \times 90/96.57 = 38.94 \, \text{mm}$$

$$y_{pl} = y_5 = y_6 - s_{pl}\cos(v) = 95 - 41.79 \times 35/96.57 = 79.86 \, \text{mm}$$

Table 10.2. Spreadsheet for the effective cross-section

Node	t	Coordinates		Effective cross-section		
		y	z	A	S_{y0}	I_{y0}
0		30	84	(J.5)	(J.7)	(J.8)
1	0.4362	35	84	2.2	183	15 389
2	0.4362	43	90	4.4	379	33 029
3	0.3897	51.5	90	3.3	298	26 831
4	0.6254	60	90	5.3	478	43 059
5	0.3021	80.09	38.34	16.7	1 075	72 676
6	0.7	95	0	28.8	552	14 110
7	0.7	140	0	31.5	0	0
			Sum	92.2	2 966	205 094

Table 10.3. Spreadsheet for the plastic section modulus

Node	t	Coordinates		Upper half		Bottom half		
		y	z	A	S_{y0}	t	A	S_{y0}
0		30	84	(J.5)	(J.7)		(J.5)	(J.7)
1	0.7	35	84	3.5	294			
2	0.7	43	90	7.0	609			
3	0.7	51.5	90	6.0	536			
4	0.7	60	90	6.0	536			
5	0.7	79.86	38.94	38.3	2472			
6		95	0			0.7	29.2	569
7		140	0			0.7	31.5	0
			Sum	60.75	4446		60.75	569

The centre of gravity for the upper and bottom parts is calculated in a spreadsheet, as in Table 10.3:

$$z_{gc,u} = \sum S_{y0,i} / \sum A_i = 4446/60.75 = 73.19 \text{ mm} \tag{J.7}$$

$$z_{gc,b} = \sum S_{y0,i} / \sum A_i = 569/60.75 = 9.37 \text{ mm} \tag{J.7}$$

The plastic section modulus is half the area times the distance between the centres of gravity for the two halves. Per unit width it is

$$W_{y,pl} = \frac{A_g/2(z_{gc,u} - z_{gc,b})}{b_{pi}/2} = \frac{60.75(73.19 - 9.37)}{220/2} = 35.24 \text{ mm}^3/\text{mm} \tag{D10.23}$$

10.6. Support reaction

Clause 6.1.7

Formulae for the resistance to the support reaction are provided in *clause 6.1.7* of EN 1999-1-4, and are entirely based on tests. For a **web without a fold**,

$$R_{w,Rd} = \alpha t^2 \frac{\sqrt{f_o E}}{\gamma_{M1}} \left(1 - 0.1\sqrt{\frac{r}{t}}\right)\left(0.5 + \sqrt{0.02\frac{l_a}{t}}\right)\left[2.4 + \left(\frac{\varphi}{90}\right)^2\right] \tag{6.12}$$

where:

r is the inner corner radius at the bend towards the support
φ is the slope of the web relative to the web
l_a is the effective bearing length
α is a coefficient for the relevant category of support.

Values of α and l_s for trapezoidal sheeting are given in Table 10.4. The values for internal supports apply if the shear forces on both sides of the support satisfy $|V_{max}| < 1.5|V_{min}|$,

Clause 6.1.7.2(5)

otherwise see *clause 6.1.7.2(5)*.

Table 10.4. Values for α and l_a for trapezoidal sheeting

End support (category 1), $c \leq 1.5h_w$	Inner support (category 2), $c > 1.5h_w$
$\alpha = 0.075$	$\alpha = 0.15$
l_a = bearing length $s_{s,}$ but < 40 mm	l_a = bearing length s_s, but <200 mm and $|V_{max}| < 1.5|V_{min}|$

Figure 10.9. (a) Support reaction and geometry of sheeting with a fold in the webs, (b) end support and (c) inner support with moment distribution

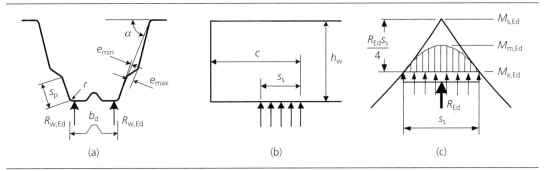

(a)   (b)   (c)

For a **web with one or two folds**, the value according to *formula 6.12* is multiplied by

$$\kappa_{a,s} = 1.45 - 0.05\frac{e_{max}}{t} \qquad \text{but not more than } 0.95 + 35\,000\,\frac{t^2 e_{min}}{b_d^2 s_p} \tag{6.16}$$

where:

b_d is the overall width of the loaded flange

e_{max}, e_{min} are the largest and smallest distances according to Figure 10.9(a)

s_p is the slant height of the plane web cross-section nearest to the loaded flange according to Figure 10.9(a).

These formulae apply on condition that the web folds are on the opposite side of the system line between the points of intersection of the midline of the web with the midlines of the flanges. The condition $2 < e_{max}/t < 12$ should also be satisfied.

10.7. Combined bending moment and support reaction

For trapezoidal sheeting, the interaction between the bending moment and the support reaction often constitutes the design criterion. The interaction formula in **clause 6.1.11** of EN 1999-1-4 for this case is empirical:

Clause 6.1.11

$$0.94\left(\frac{M_{Ed}}{M_{c,Rd}}\right)^2 + \left(\frac{R_{Ed}}{R_{w,Rd}}\right)^2 \leq 1 \tag{6.22}$$

In *expression 6.22*, the bending moment M_{Ed} may be calculated at the edge of the support, $M_{e,Ed}$ (see Figure 10.9(c)). If it is assumed that the reaction force is uniformly distributed over the bearing length s_s, the theoretical maximum moment $M_{s,Ed}$ over the support (determined by assuming that the support bearing length = 0) is reduced by $R_{Ed}s_s/4$. At the centre of the support the moment is reduced by half that value, so the moment resistance of the sheeting should be checked for the moment $M_{m,Ed} = M_{s,Ed} - R_{Ed}s_s/8$.

10.8. Flange curling

EN 1999-1-4 **clause 5.4** states that the effect of flange curling on the load-bearing resistance should be taken into account when the magnitude of curling (inward curvature towards the neutral axis) is greater than 5% of the depth of the cross-section. For initially straight beams, *expression 5.1e*, which applies to both the compression and the tension flanges, with or without stiffeners, is provided. For arched sheeting, where the curvature, and therefore the force components on the flanges, is larger, *expression 5.1f* is provided.

Clause 5.4

$$u = \frac{2\sigma_a^2 b_s^4}{E^2 t^2 z} \tag{5.1e}$$

$$u = \frac{2\sigma_a b_s^4}{E\, t^2 r} \tag{5.1f}$$

where:

u is the magnitude of flange curling towards the neutral axis

b_s is half the distance between the webs

z is the distance from the flange under consideration to the neutral axis

r is the radius of curvature of an arched sheeting

σ_a is the mean stress in the flange (if the stress is calculated for the effective cross-section, the mean stress is obtained by multiplying the stress for the effective cross-section by the ratio of the effective flange area to the gross flange area).

If the magnitude of flange curling is found to be greater than 5% of the depth of the cross-section, then a reduction in load bearing resistance, due to, for instance, the reduction in depth of the section or to possible bending of the webs, should be made.

In order to avoid distortion of the trapezoidal cross-section during erection, the inclination of the web defined by the angle φ in Figure 10.9(a) should not be less than 65°.

10.9. Other items in EN 1999-1-4

10.9.1 Shear force

In trapezoidal sheeting, the shear force and the combined moment and shear force constitute the design criterion only if web crippling is prevented by support reinforcement. Resistance to shear force is in principle the same as for extruded or welded profiles. See **Clauses 6.1.5** and **6.1.10** of EN 1999-1-4.

Clause 6.1.5
Clause 6.1.10

10.9.2 Compression

Resistance to compression is covered in *clause 6.1.3* of EN 1999-1-4, and combined compression and bending in *clause 6.1.9*.

Clause 6.1.3
Clause 6.1.9

10.9.3 Shear lag

In *clause 6.1.4.3* of EN 1999-1-4 it is stated that the effects of shear lag should be taken into account according to EN 1999-1-1. Shear lag effects may be ignored for flanges with $b/t \leq 300$.

Clause 6.1.4.3

10.9.4 Stressed skin design

Diaphragms may be formed from structural sheeting on roofs. Some overall conditions for stressed skin design are given in *clause 6.3* of EN 1999-1-4. Further information on the verification of such diaphragms can be obtained from Baehre and Wolfram (1986) and ECCS Publication No. 88 (ECCS, 1995).

Clause 6.3

Table 10.5. Fastener material with regard to the corrosion environment. Only the risk of corrosion is considered. Environmental corrosivity categories according to EN ISO 12944-2.

Corrosivity category	Sheet material	Material of fastener					
		Aluminium	Electro-galvanised steel: coat thickness $\geq 7\ \mu m$	Hot-dip zinc-coated steel:[b] coat thickness $\geq 45\ \mu m$	Stainless steel, case hardened: 1.4006[d,e]	Stainless steel: 1.4301[d] 1.4436[d]	Monel[a]
C1	A	X	X	X	X	X	X
C2	A	X	–	X	X	X	X
C3	A	X	–	X	–	X	X
C4	A	X	–	(X)[c]	–	(X)[c]	–
C5-I	A	X	–	–	–	(X)[c]	–
C5-M	A	X	–	–	–	(X)[c]	–

Key: A, Aluminium irrespective of surface finish; X, Type of material recommended from the corrosion standpoint; (X), Type of material recommended from the corrosion standpoint under the specified condition only; –, Type of material not recommended from the corrosion standpoint.
[a] Refers to rivets only. [b] Refers to screws and nuts only. [c] Insulation washer of a material resistant to ageing between the sheeting and fastener. [d] Stainless steel EN 10088. [e] Risk of discoloration.

Table 10.6. Atmospheric corrosivity categories according to EN ISO 12944-2 and examples of typical environments

Corrosivity category	Corrosivity level	Example of typical environments in temperature climate (informative)	
		Exterior	Interior
C1	Very low	–	Heated buildings with clean atmospheres, e.g. offices, shops, schools and hotels
C2	Low	Atmospheres with a low level of pollution. Mostly rural areas	Unheated buildings where condensation may occur, e.g. depots and sport halls
C3	Medium	Urban and industrial atmospheres, moderate sulphur dioxide pollution. Coastal areas with low salinity	Production rooms with high humidity and some air pollution, e.g. food-processing, plants, laundries, breweries and dairies.
C4	High	Industrial areas and coastal areas with moderate salinity	Chemical plants, swimming pools, and coastal shipyards and boatyards.
C5-I	Very high (industrial)	Industrial areas with high humidity and aggressive atmospheres	Buildings and areas with almost permanent condensation and with high pollution
C5-M	Very high (marine)	Coastal and offshore areas with high salinity	Buildings and areas with almost permanent condensation and with high pollution

10.9.5 Perforated sheeting
Formulae for the resistance of perforated sheeting with the holes arranged in the shape of equilateral triangles are given in *clause 6.4* of EN 1999-1-4.

Clause 6.4

10.9.6 Serviceability limit state
For the calculation of deformations at the serviceability limit state, see Chapter 7 in this guide.

10.9.7 Testing procedures
Procedures for the testing of profiled sheeting are given in *Annex A* of EN 1999-1-4.

10.10. Durability of fasteners
For the basic requirements on the durability of aluminium structures, see *Section 4* of EN 1999-1-1. For cold-formed structural sheeting, special attention should be given to the risk of corrosion of mechanical fasteners. Recommendations for the choice of fasteners for the environmental corrosivity categories defined in EN ISO 12944-2 are given in *Annex B* of EN 1999-1-4, and the data are partly reproduced in Table 10.5.

The environmental corrosivity categories following EN ISO 12944-2 are presented in Table 10.6.

REFERENCES
Baehre R and Wolfram R (1986) *Zur Schubfeldberechnung von Trapezprofilen.* Stahlbau 6/1986, S. 175–179.
ECCS (1995) *European Recommendations for the Application of Metal Sheeting Acting as a Diaphragm.* European Convention for Constructional Steelwork, Brussels. ECCS Publication No. 88
Höglund T (1980) *Design of Trapezoidal Sheeting with Stiffeners in the Flanges and Webs.* Swedish Council for Building Research D28, Stockholm.

Designers' Guide to Eurocode 9: Design of Aluminium Structures
ISBN 978-0-7277-5737-1

ICE Publishing: All rights reserved
http://dx.doi.org/10.1680/das.57371.197

Index

Page locators in *italics* refer to figures separate from the corresponding text.